About Island Press

Since 1984, the nonprofit organization Island Press has been stimulating, shaping, and communicating ideas that are essential for solving environmental problems worldwide. With more than 800 titles in print and some 40 new releases each year, we are the nation's leading publisher on environmental issues. We identify innovative thinkers and emerging trends in the environmental field. We work with world-renowned experts and authors to develop cross-disciplinary solutions to environmental challenges.

Island Press designs and executes educational campaigns in conjunction with our authors to communicate their critical messages in print, in person, and online using the latest technologies, innovative programs, and the media. Our goal is to reach targeted audiences—scientists, policymakers, environmental advocates, urban planners, the media, and concerned citizens— with information that can be used to create the framework for long-term ecological health and human well-being.

Island Press gratefully acknowledges major support of our work by The Agua Fund, The Andrew W. Mellon Foundation, Betsy & Jesse Fink Foundation, The Bobolink Foundation, The Curtis and Edith Munson Foundation, Forrest C. and Frances H. Lattner Foundation, G.O. Forward Fund of the Saint Paul Foundation, Gordon and Betty Moore Foundation, The JPB Foundation, The Kresge Foundation, The Margaret A. Cargill Foundation, New Mexico Water Initiative, a project of Hanuman Foundation, The Overbrook Foundation, The S.D. Bechtel, Jr. Foundation, The Summit Charitable Foundation, Inc., V. Kann Rasmussen Foundation, The Wallace Alexander Gerbode Foundation, and other generous supporters.

The opinions expressed in this book are those of the author(s) and do not necessarily reflect the views of our supporters.

CHASING THE RED QUEEN

Chasing the Red Queen

THE EVOLUTIONARY RACE BETWEEN
AGRICULTURAL PESTS AND POISONS

Andy Dyer

Washington | Covelo | London

ISLAND PRESS is a trademark of the Center for Resource Economics.

Library of Congress Control Number: 2014942401

Printed on recycled, acid-free paper ⊕

Manufactured in the United States of America

10 9 8 7 6 5 4 3 2 1

Keywords: Island Press, pesticide resistance, microbial ecology, sustainable agriculture, eco-agriculture, monoculture, trophic cascades, biotechnology, genetics, Darwin, Roundup

"Well, in our country," said Alice, still panting a little, "you'd generally get to somewhere else—if you ran very fast for a long time, as we've been doing."

"A slow sort of country!" said the Queen. "Now, here, you see, it takes all of the running you can do to keep in the same place. If you want to get somewhere else, you must run at least twice as fast as that!"

—*Lewis Carroll,* Through the Looking-Glass, *1871*

CONTENTS

Our mental image of a farm is a peaceful setting with fields of wheat or corn gently waving in the afternoon breeze and a farmer driving a tractor, patiently tending the crops until they are ready for harvest. This tranquil scene comes not from personal contact with agriculture, which most of us don't have, but primarily through the imagination and exaggeration of Hollywood. In truth, farming is very hard work with long days that often begin before dawn and end after dark. In contrast to the bucolic Hollywood scenes, the farm is often more like a battlefield, with the farmer spending much of his or her time and energy in a war with nature.

The enemies are insects, mites, bacteria, fungi, nematodes, and any number of microbes that chew, suck, tunnel, infect, and otherwise reduce the vitality of the crop plants. This battle began about 10,000 years ago with the dawn of agriculture, and it takes place, in some form, in every human culture. But today, farmers are losing their foothold—and the war is growing ever more expensive. It is increasingly difficult to farm economically on a small scale; feeding the world has become an enormous industrial enterprise, practiced intensively, with ever more advanced equipment and technology.

Part of the problem is that we have been remarkably success-ful at making our food crops desirable not just to us, but to many, many other species. We've ramped up the sweetness and juiciness of fruits, the quantity of starches in grains, the proportions of proteins and fats in seeds, the overall yield of crops. Many of today's fruits and grains bear faint resemblance to their wild progenitors. We find ourselves fending off the rest of nature because we usually grow these tasty plants in vast single-species stands and make every attempt to eliminate all other competing species in the fields. Like us, the many species of crop pests are eager to take advantage of these incredibly abundant, incredibly palatable, and increasingly vulnerable resources.

To combat these pests, the weapons of agriculture have grown increasingly lethal, from elemental compounds and plant extracts a century ago to potent synthetic herbicides and pesticides today. Yet modern farmers have no more success against weeds and other invaders than their ancestors did several generations ago. And in addition to being more expensive, synthetic pesticides are more damaging to the environment. Each growing season, farmers use the latest chemicals and technology to protect their crops, yet the goal of long-term control of crop pests remains elusive. Instead, the pests simply adapt, becoming increasingly resilient to an increasingly costly succession of pesticides. In addition, there is tremendous collateral damage to many other organisms in and around the fields and in the soil; many of these organisms are necessary for their contributions to the health of the environment. This collateral damage has important and long-term effects on our ability to maintain farm productivity.

The constant struggle to control crop losses can be described in military terms as a battle, but an even better analogy comes from Lewis Carroll's *Through the Looking-Glass*.[1] The Red Queen explains to Alice that, in Wonderland, a person must run very fast just to stay in the same place—a metaphor that biologists have borrowed to explain the process of coevolution. In biology, the Red Queen Hypothesis describes how an organism adapts to an environmental stress, such as the actions of a second organism, which elicits a counteradaptation in the second organism. The result is that both sides are continuously adapting and counteradapting to each other. In such an "evolutionary arms race" there is no winner, only a never-ending race without a finish line.

This is the situation in which farmers currently find themselves. As new pesticides are produced and applied to kill unwanted organisms, the targeted pests adapt to each new chemical, which requires the development of new chemicals, which stimulates further adaptation, and so on. For 60 years, farmers have become increasingly entrapped by what has been termed a "chemical dependency treadmill." To break the cycle, we must first recognize the biology behind it. This book will explore the central problem of modern agriculture—pest resistance—by applying the principles of evolutionary biology. We will use the Red Queen Hypothesis as a mental construct for understanding the way species and populations reduce environmental stress by adapting to it. More importantly, the Red Queen Hypothesis will help us understand why the use of chemical controls has no real end point and why there is no realistic chance of winning the war against crop pests.

Most of us have at least some familiarity with efforts to control, and even eradicate, the creatures who are trying relentlessly to eat our food before we can. Those efforts are the result of billions of dollars spent on research and development by agrochemical companies to provide farmers and ranchers with the chemical tools they need to kill agricultural pests, including weeds, insects, and disease-causing pathogens. We can also appreciate that chemical companies, like any other business, are economic competitors and must produce better and different chemicals as part of their effort to maintain market share and corporate earnings. What many of us may not realize is that evolutionary biology drives the entire process and restricts the success of any and all efforts at pest control. The rules of engagement, essentially the rules of the evolutionary war, will be described in this book.

This conversation extends well beyond our attempts to control pests and must also focus on protecting soil and biodiversity, and on sustainable farming such that we are able to maintain our capacity to produce food into the future. While agriculture has always been the backbone of the American economy, it is now producing goods for a growing range of sectors of the economy completely unrelated to food. For example, corn (maize) has always been a vital commodity in American agriculture, but the production of ethanol is now the largest "consumer" of corn in America, and additional products derived from corn include plastics, explosives, insecticides, adhesives, dyes, construction materials, paints, and paper products. Corn production now has a life of its own beyond providing food for humans and livestock, and our economy is increasingly reliant on this single crop. Therefore, any threat to current levels of farm production has far-reaching consequences, and protecting that productivity by any means possible has become a national priority. It is perhaps ironic that we may ultimately have less to fear from terrorism or trade barriers or market manipulation in agriculture than we do from multitudes of tiny organisms that are completely indifferent to our economic status or national goals. And we have to be concerned because we have been largely unsuccessful at dictating the terms of this battle.

The no-win scenario of the Red Queen has not been applied to chemical pest control in agriculture even though chemical resistance and the evolutionary process behind it are very well known and have been documented in everything from bacteria and fungi to mosquitoes and nematodes. However, as our technological capacity has expanded, our creative attempts to control pests have also expanded. Today, an entire industry is oriented around developing genetically modified plants that can withstand the toxic effect of specific herbicides, yet we are already faced with the emergence of weeds that are resistant to those herbicides. Thus, despite our apparent understanding of evolutionary biology, we are failing to recognize that agriculture is not exempt from the same rules that govern all living things. My goal with this book is to convince the reader that, despite the enormous commitment of time, money, and expertise to the development of chemical and biochemical approaches in pest control, the race with the Red Queen is not a race that can be won.

The Red Queen Hypothesis offers a simple yet profound tool for helping us understand how nature works in relation to the problems we continually encounter in agriculture. Just as important, this understanding also offers us a direction for how we can move from fighting nature to benefiting from it in the business of food and fiber production. Increasingly, we see that adhering to the rules of ecology and evolutionary biology may be the key to managing agricultural productivity. This book provides a few historical and contemporary examples in which farmers and resource managers work with natural processes to strengthen agriculture by integrating crop plants into a healthier farm ecosystem. By doing so, they reduce or even eliminate the need for artificial chemicals, whether pesticides or fertilizers, and regain a more natural control over crop pests.

The concept of the Red Queen concept is simple, but I want to acknowledge that the problem of pesticide resistance is complex. The scope of this book does not allow me to explore all the intricacies of the concepts I discuss, and I hope that knowledgeable readers will excuse any simplifications. After asking colleagues from different disciplines in biology to review the text of this book, I recognize that my views may differ from those of some others, but that's part and parcel of the process of science. I would like to gratefully acknowledge friends and colleagues for constructive comments on this book as it was being developed: Kevin Rice, Buz Kloot, Ray Archuleta, Nathan Hancock, Michele Harmon, Hugh Hanlin, Garriet Smith, Bill Jackson, and Derek Zelmer, as well as my students Alyssa Smith and Brandy Bossle.

PART I

Introducing the Red Queen

In the game of life, less diversity means fewer options for change. Wild or domesticated, panda or pea, adaptation is the requirement for survival.

—Cary Fowler

Chapter 1

The Never-Ending Race: Adaptation and Environmental Stress

In the natural world as well as the business world, staying one step ahead is the key to success. However, it isn't possible for everyone to stay one step ahead of everyone else. One is reminded of Garrison Keillor's Lake Wobegon, where "all of the children are above average." Clearly, it isn't possible for everyone to win the race for success—and it *is* a race. In nature, regardless of the particular situation, those that fall behind become food for others. The Red Queen's advice to Alice suggests to us that adapting to an ever-changing world is a continuous requirement for survival; being good isn't enough, and we must work constantly to stay ahead of the pursuers . . . and the competitors . . . and the predators.

An analogy we can use is that of the fox, a predator, and rabbits, the fox's prey. The fox pursues, the rabbits run. If the fox catches the slowest rabbits, then only the fastest rabbits remain in the population and their offspring (the next generation) should be faster than the average rabbit of the present generation. This is the basis for the phrase "survival of the fittest": those individuals that are most "fit" in this environment are most likely to survive to reproduce.[1] However, as the prey population becomes faster and better able to avoid the predators, the predator population will die out unless it adapts to be fast enough to continue to catch the slower prey.[2] Hence, the faster and more successful foxes produce faster fox pups, while the slower foxes fail to survive or reproduce. Logically, this process of adaptation appears to be a positive feedback cycle. If the foxes are continually catching the slowest rabbits, the rabbit population will become faster and faster over time until we witness supersonic rabbits flashing around being chased by equally supersonic foxes. Obviously, this *reductio ad absurdum* result does not happen, and it is important to understand why.

First, there are limits to how fast a rabbit (or a fox) can run. Even if the genetic potential existed in the population, the energy requirements, the physiological demands, and the physical properties of the body all interact such that there are limits to the range of possible modifications. A supersonic rabbit would be all legs, with an incredibly high metabolism, and bones and sinews made of something unusually strong. Such a rabbit certainly wouldn't eat grass.

Second, and more importantly, running faster isn't the only solution to fox predation for the rabbits. Hiding, camouflage, mimicry, early detection, evasive action, changes in activity times, movement to predator-free habitats, claws and teeth, toxins, and group defense are all examples of adaptations used by animals to avoid or prevent predation. A population lacking adaptive options is a population that will soon run its course.

Regardless of the mechanism, the Red Queen demands that a population adapt or it will fall behind in the race for survival. Adapting isn't optional; it is mandatory: adapt or go extinct. The fittest individuals survive, but the definition of "fittest" can change with every generation as the conditions change. Therefore, for every stress or challenge or demand in the environment, organisms that respond in an appropriate way are more likely to survive than those that do not respond appropriately. The challenges of the environment are many and varied and may not be the same from one year to the next, but the challenge to the individual is the same: meet the demands of the environment or become food for those that do.

Simply put, all adaptations are a response to environmental stress. A "stressor" can be thought of as any influence in the environment that lowers the ability of an individual in a population to survive and reproduce. An adaptation is a trait that reduces the negative effects of that stress. To be clear, adaptations do not *eliminate* stress; they only *reduce* the stress experienced by the individuals in that particular generation. Those individuals able to tolerate an environmental stress are more likely to live longer than those less tolerant, and as a result are more likely to produce more offspring. Therefore, the survival of a population or a species is a process of responding to stress, and because there are myriad different potential stressors in the environment, this process is constant and ongoing: it never ends. Nonetheless, however challenging survival may be to each individual in a population, it is only necessary to stay one step ahead of the pack. As the saying goes, when you're being chased by a bear, you don't have to run faster than the bear, just faster than the person next to you.

~

The premise of the Red Queen has been adopted by evolutionary biologists to exemplify the concept of continual adaptation to the constancy of environmental stress. Specifically, the Red Queen Hypothesis was originally used to understand the tight relationship that can evolve between a pathogen (or parasite) and its host. In such a situation, the pathogen must be successful at defending itself from the protective measures of the host, but the pathogen cannot be overly successful in a numerical sense or it will kill the host. If the host dies, the pathogen dies unless it can successfully transfer to a new host. Any host that can defend itself from the pathogen will be much more successful than those that can't, but any pathogen that succumbs to the defensive measures of the host will be replaced by those that can resist. Thus, over the long term, a sort of détente evolves wherein pathogens are successful enough to persist and the hosts are successful enough not to die *too quickly*. In both cases, success can be measured as "lives long enough to reproduce"—or to infect another host, in the case of the pathogen. For both, their adaptations for survival allow the environmental stress to be reduced, but not eliminated, and the race goes on.

One critical component in this process is this: all adaptations are a function of time and this time is measured in generations. Ultimately, the only reason any species has ever become extinct is that the stress the species experienced was operating on a shorter time scale than the adaptive process could accommodate.[3] The Irish elk, the largest deer that ever lived, did not die out because of an inability to adapt, but because the changes to the environment at the end of the last ice age occurred faster than the Irish elk could adapt to them.[4] Obviously, this huge deer could and did adapt—males could be seven feet tall at the shoulders with twelve-foot-wide antlers—but the changes in the post-glacial environment of Europe and western Asia occurred more quickly than the giant deer could cope with. It is certainly possible that expanding human populations and their technology may have compounded the stress.

Similarly, the extinctions of the great auk, the passenger pigeon, and the sea mink did not occur because there are no possible adaptations to the activities of humans, but because the time needed for such adaptations is longer than the amount of time the species had available to them.[5] The great auk, for example, was not able to tolerate the simultaneous stresses of human hunting and egg collecting. What was different about those stresses compared to those the great auk had faced for thousands of years prior? Do humans create greater stress than ice ages or polar bear predation?

The answers lie in an understanding of the principles of evolutionary biology. Those populations that can adapt quickly can stay one step ahead; they can successfully respond to the Red Queen's admonition to "run faster." The stronger and more intense the stressor, the faster the adaptive process must operate to reduce the stress and, consequently, the more likely it is that populations with slow adaptive responses will fail to adapt. Failing to adapt even once means extinction, and those species with slow response times are therefore more likely to become extinct. And it should come as no surprise that far more extinctions are seen in large organisms than in small organisms. Why are large organisms so slow to adapt? Why does it make any difference what the size of the organism is when it comes to responding to environmental stress? Why are there more insects than tigers?

As a general rule, when the individuals of a population encounter a new environmental stress, some individuals in the population will die prematurely and some individuals will survive. If the reason for their survival (such as a slightly enhanced ability or trait) can be passed on to their offspring (that is, it's genetically encoded), then the next generation should be better able to withstand the newly encountered environmental stress, and the population overall will be less susceptible to it. Therefore, the first key to the ability of a population to survive by adaptation lies in the rapidity with which the next generation, the more resistant generation, is produced by the survivors of this generation. It follows that those species capable of producing a new generation very quickly should be better able to respond to a stress very quickly. Those species that require more time for reproduction will be slower to adapt to the stress because of the additional time needed for them to produce stress-tolerant offspring.

For example, consider three very different organisms: bacteria, houseflies, and elephants. A bacterium can reproduce every 20 minutes. A housefly may lay up to 500 eggs, and the offspring can be laying eggs of their own in as little as a week. Elephants can produce a baby every four years, and the offspring may require 10–15 years to mature to the point where they can produce a single baby of their own. One bacterium reproducing every 20 minutes can (potentially) produce 72 generations of ~5 x 10^{21} (5 sextillion) descendants in 24 hours. The offspring of one housefly can (potentially) produce four generations numbering 4 x 10^9 (4 billion) individuals within a month. One elephant can (potentially) produce six young in 30 years. Using similar parameters, Charles Darwin calculated that elephants would need 750 years to produce 19 million individuals. Bacteria, houseflies, and elephants *do not* adapt at the same rate.

The production of large numbers of offspring is not a survival necessity, but high reproductive output is definitely correlated with the very short generation time that is critical to rapid adaptation. The housefly produces four generations in a month and those offspring can soon number in the billions. However, the fact that we are not (usually) overrun with houseflies indicates that organisms that reproduce at very high rates and in large numbers also experience very high mortality rates.

A second factor in survival via rapid adaptation to environmental stress is the size of the population. As a rule, the larger the population, the more likely the species will be able to cope with environmental stress because of the greater amount of genetic variation. A population of a billion houseflies is far more likely to contain a wider range of genetic variation and, therefore, of stress-tolerant individuals than a population of a hundred houseflies. And if those houseflies are descendants of a partly or wholly stress-tolerant individual, the *alleles* of the *gene* (see box 1-1) for that tolerance are likely to exist in large numbers, too. A contrasting but equally well-established principle in ecology is that physically large organisms have lower population densities than do smaller organisms. This general relationship, again, makes it more likely that smaller-bodied organisms, with rapid generation times and larger population sizes, have the greater capacity to adapt more quickly than large-bodied organisms.

A third factor controlling adaptation in organisms is that the intensity of the environmental stress will influence the rate at which the population can adapt. We assume that only a small proportion of a population is likely to be tolerant of a novel stress in the environment. If a low-intensity stress kills only a few individuals, the remaining population will comprise individuals across the entire range of tolerance. While the genetic variation within the population may have been reduced, the offspring in the next generation will still represent a wide range of genetic variation. However, if a high-intensity stress kills a large majority of the population, only the most tolerant individuals will remain. The resulting offspring will represent only the very narrow range of genetic variation that

Box 1-1: All the genetics you'll need

Throughout this book there will be references to the genetic makeup of individuals and populations. For now, consider all species to have two copies of each *chromosome,* just as humans do. One set is the maternal contribution and one set is the paternal contribution to their offspring. A specific chromosome contains a number of *genes,* and the genes are the DNA code for specific proteins. All individuals in a species possess the same genes, but each individual is likely to have different versions of many of those genes. These versions are called *alleles.* Your parents most likely gave you different alleles for each of the genes on each of your 23 different chromosomes. Only very closely related individuals are likely to have many of the same alleles.

When the two alleles are the same, an individual is *homozygous* for that gene, but *heterozygous* if the two alleles are different. A *dominant* allele will mask a *recessive* allele (in a heterozygous individual) and the expression of the recessive alleles will often only be seen when there are two copies in a homozygous individual. For example, in humans, albinism is the absence of pigmentation (melanin) and is a recessive trait only seen in individuals who have two copies of that allele. Heterozygotes with only one copy of the dominant allele appear normally pigmented because the dominant allele allows normal levels of melanin to be produced.

A change in the genetic code (DNA) of a normal allele results in a *mutation.* Mutations are random errors in the copying (transcription) of the DNA. (If a mutation in an important gene has a very negative effect, the individual will likely not survive and the mutation will be eliminated.) If mutations occur in sperm or egg cells, they can be passed on to the offspring that result from that particular sperm or egg. Most mutations have no strong effects, but some have a negative effect and a few will have a positive effect. In very large populations (say, a billion houseflies), there are almost certainly a large number of mutations present, and some of these may be positive and increase the survival chances of the possessors of the mutation. For a mutation to spread in a population, the environment must favor the possessor of the mutation in some way, and that advantage must lead to a relatively higher reproduction rate of that individual compared to the rest of the population.

confers resistance to the environmental stress. Given a sufficiently intense stress, with all of the offspring being descendants of very tolerant survivors, adaptation could occur as quickly as one generation. The population will be small, but completely tolerant of the stress.

Taken together, the interactions among reproduction rate, population size (and therefore genetic variation), and the intensity of stressors helps us to understand the response and survival of populations. A stress that kills a large majority of the individuals is more problematic for a small population because the subsequent recovery will be based on the few remaining individuals, and the available genetic variation in small populations is restricted. Thus, when populations are small, an adaptive response to stress is more problematic. However, large populations are less susceptible to extinction from intense stressors because of their sheer size and the greater likelihood of containing stress-resistant genetic variation. If 90 percent of a large population is lost, the survivors still represent a large number of individuals and a large amount of genetic variation. Thus, for large populations, the likelihood of extinction is lower, the likelihood of having resistant individuals in the population is higher, and the likelihood of surviving an intense stress is greater. The implications for dealing with very large populations of problematic species, such as pests, pathogens, and invasive species, should be obvious.

Once a population has experienced significant mortality from an environmental stress, the time needed for the population to recover will depend on the generation time and the number of offspring produced. The recovery may require the *same number* of generations for both small and large organisms, but that recovery will occur over a very short period of time for small organisms with short generation times and over a much longer time for larger organisms with long generation times. And, of course, if the small organisms produce much larger numbers of offspring each generation, those populations will also recover more quickly in the numerical sense.[6]

Later, we will consider two additional factors that influence the process of adaptation: the duration of the stress and the spatial scale of the stress experienced by a population. As before, if an extreme stress eliminates the majority of individuals in a small area, only stress-resistant individuals (if any) are likely to remain. However, if the population in the areas adjacent to the stress is sufficiently large and healthy, those individuals can easily move into the now unoccupied space. If the stress has abated, those individuals do not need to be resistant; they can mix with the survivors remaining in the previously stressed habitat, and the population can survive into the future with no particular need for adaptation. This immigration of outside individuals into the small population of survivors, termed "gene flow," will dilute the adaptive response to the environmental stress. If, however, the environmental stress is sufficiently widespread or long-lasting, the evolutionary response is different. The time needed for gene flow to affect the population of survivors will increase as the spatial scale of the stress increases, and this will slow the dilution effect, particularly if individuals are not able to disperse quickly or across great distances. Similarly, if the stress persists across multiple generations, then nonresistant immigrants from unaffected areas will not survive and will not affect the genetic composition of the surviving population.

Overall, when stressful conditions exist over large areas, we see that small organisms that reproduce quickly with short generation times can adhere to the dictates of the Red Queen far more readily than large organisms. For a large population of small, rapidly reproducing organisms, even a very severe stress that kills the vast majority of individuals is survivable. If only a few individuals survive, the population can still recover, and the subsequent generations will be largely resistant to the stress that killed most of the original generation. In fact, if resistant individuals occur naturally in a population, then the greater the proportion of individuals that are killed by the stress, the faster the species will adapt to that stress because only the resistant individuals will survive and reproduce. Therefore, though it may seem a paradox, for large populations of rapidly reproducing organisms, *the more powerful and intense the lethal stress, the faster the development of resistance to it.* This will become an important theme in later chapters.

Box 1-2: The scale of evolution

Evolution is the change in the frequency of an allele (or alleles) within a population from one generation to the next. If an environmental stress favors a particular trait, the alleles for that trait are likely to be more common in the future because individuals possessing that trait produce offspring with that trait, and the population changes in terms of its genetic composition. In other words, the effects of natural selection are such that individuals live and die because of the traits they possess, and the differential reproduction of the survivors changes the allelic frequencies in the population. This change in the genetic composition of a population such that more individuals possess a specific trait that allows them to survive better is called *adaptation*. The population shows a genetic shift in response to the stress that caused certain individuals to die and, in this way, the surviving population adapts.

In contrast, individuals cannot evolve and they do not acquire new adaptations, although they are capable of adjusting or "acclimating" to changes in the environment. Every individual is born with the particular DNA of its species: a set of chromosomes containing all of the genes (but not all of the alleles) for every trait the species possesses. At no point in its lifetime will an individual gain any more or different alleles for those genes. The complete genetic code for the individual is contained in every cell of the body, and it does not change during the life of the individual. If a mutation were to occur in an individual, it would occur in a single cell and would not affect the rest of the body. Even if the mutation proliferated as the original cell divided, it would still be very localized (often resulting in very negative effects such as cancer) and could not become part of every cell in the body.[a]

The consequence of having fixed genetics is that individuals are not capable of acquiring new adaptations, but capable only of adjusting physiologically to their environment over a season or a lifetime. This process of an individual adjusting physiologically to response to stress is called acclimation. The incorrect use of the term *adapt* to mean *acclimate* is a frequent problem when discussing topics in evolutionary biology, and it is very important to understand the difference between the two concepts.

---

a. Sadly, this rules out any prospect of gaining superpowers as a result of irradiation or alchemy.

~

With that understanding, we now have a way to interpret a remarkable series of events in recent human history. Since World War II, hundreds of organisms have demonstrated their ability to obey the Red Queen and to repeatedly develop resistance to the environmental stresses imposed by humans. Adaptations are frequently noted in undesirable organisms such as viruses and bacteria, microbes such as malaria, fungi such as candida, and insects such as houseflies and mosquitoes. What changed after WWII was our ability to attack unwanted organisms following our development of more potent medicines, discoveries of antibiotics, and the production of synthetic pesticides. Since the introduction and widespread use of such modern chemicals, we have seen the emergence of the great majority of resistant organisms.

In no other sector of human activity has this process of adaptation to chemicals been more evident than in agriculture. In fact, the evidence is so comprehensive that modern farming could be considered a model for observing evolutionary biology in action. Every intense and large-scale attempt to eradicate an agricultural pest has been met with an adaptive response by the pest, which has generated modified efforts by humans to eradicate the pest, which has in turn engendered new adaptive responses by the pest, and so on. For every move there is a countermove as both people and pests try to "run at least twice as fast as that." The Red Queen has defined the rules of a game for which there does not seem to be a winning strategy.

Chapter 2

The Evolution of Farming: Scaling Up Productivity

Long ago, farming was done at a small scale, was oriented around single families, and focused on multiple crops. The goal was growing food for the family and, in good years, for selling and bartering any surplus. Production was limited by muscle power that was provided by humans with hand tools and by animals pulling rudimentary equipment. Harvesting was a manpower issue; one could not reasonably harvest more crops than the family was capable of maintaining and storing in the time allotted. More recently, farming was facilitated by motorized equipment, such as tractors, to pull implements for cultivating the soil and harvesting the crops. As the technology advanced the capacity to farm, larger tracts of land were cultivated. Most famously, the John Deere plow allowed farmers to cultivate more land, more quickly than ever before, and thus it ushered in a new era of large-scale farming.[1] However, as that ability to expand operations became possible, the objectives of farming changed . . . and the nature of the farm itself changed, as did the identity of the farmer.

The Advent of Farming

From the advent of agriculture more than 10,000 years ago until perhaps 60 years ago, agriculture and the production of food and fiber revolved around the ability to mitigate the negative effects of climate on growth and production. Temperature and water were the two environmental variables that most influenced farm productivity, just as they determine the productivity of most natural ecosystems. Farming prospered in regions where the temperatures were moderate enough to prevent undue physiological stress on plants or where that stress could be ameliorated by the addition of water. Therefore, agriculture developed most rapidly in temperate zones where rainfall was seasonal, but predictable, and in hotter regions where water could be applied through irrigation. And because technology was primitive, unavailable, or very expensive, early agriculture was largely limited to those regions where either moderate temperatures or available water allowed a sufficiently long growing season for crops to mature.

Over time, many human societies made a transition from hunting and gathering to pastoral or farming cultures, and food production by cultivation became an increasingly important economic enterprise. However, the process remained a small-scale family endeavor, largely because it was labor intensive and technological advancements were slow to develop. Growing plants for food was limited to those species that were regionally adapted and hardy, and that required relatively low inputs from humans in order to survive, grow, and reproduce. In time, more plant species were incorporated, the techniques for raising them became more complex and advanced, and plant breeding provided more cultivars that could be raised outside of their natural climatic zones. As people collected seeds to be sown in the next growing season, they selected plants with the best traits, such as hardiness, productivity, or stress tolerance. Eventually, as technological advancements such as plows were combined with the use of draft animals, growing food crops expanded beyond the scale of the family farm and became a major global activity.

Box 2-1: Patches, pests, and time lags

Predators are able to control the abundance of prey species, but only if they can find them. As a rule, the larger the prey population, the faster the rate of discovery by predators, but this also depends on the distance between the predator and the prey population. A classic example of the dynamics between predators, prey, and patch size is that of the invasion of eastern Australia by species of prickly pear cactus (*Opuntia sp.*) and subsequent control, but not eradication, by the cactus moth (*Cactoblastis cactorum*). The cacti were introduced about 1840 as ornamental species but escaped into the New South Wales and Queensland landscape, where they found a very suitable new home. By 1870, cacti were becoming enough of a problem that federal Prickly-Pear Destruction Acts were passed in 1886, 1901, and 1924. By 1920, cactus had converted 25 million hectares of a primarily sheep-grazing region to land considered almost worthless because of dense stands of one- to two-meter-tall cactus. The infestation was so bad that the Australian government created a Prickly-Pear Destruction Commission to develop and implement eradication procedures.[a] Eventually, the *Cactoblastis* moth, whose larvae eat prickly pear cactus, was imported from the United States and its introduction to Australia is considered one of the landmark examples of biological control.[b] Within four years, the large expanses of cactus were reduced to rotting remains as the moth spread across the entire region.

It is important to note that biological control is not biological eradication, despite the intentions of those involved in the effort. Control means that the growth and expansion of the pest have been checked, but it never means that the pest has been eliminated from the landscape. In fact, the success of the control agent acts in the long run to check its own growth as well as that of the prey species. In this case, the moth nearly completely obliterated the population of cactus, but in so doing it eliminated its own food source. As the population of cactus was diminished, the size of the moth population necessarily followed suit. However, here and there, individual cacti survived the attack and, once the moths had disappeared locally due to lack of food, the cacti began to grow back.

---

a. "Prickly Pear History," North West Weeds website, 2014, www.northwestweeds.com.au /prickly_pear_history.htm.

b. H. Zimmermann, S. Bloem, and H. Klein, "Biology, History, Threat, Surveillance, and Control of the Cactus Moth, *Cactoblastic cactorum*," Joint FAO/IAEA Programme of Nuclear Techniques in Food and Agriculture (Vienna: International Atomic Energy Agency, 2004), ISBN 92-0-108304-1.

From this point on and to this day, there exists a balance between the cactus and the moth populations. As the cacti grow back, the small local patches eventually become large enough to be discovered by the moths, which then capitalize on the new food source. The process of discovery by the moths is a function of time, size, and distance.[c] From the perspective of the moths, the cacti can be viewed as small resource islands in an ocean of empty and unusable habitat. The probability that a moth will discover an island of cacti depends on the size of the island and the distance the moth must travel. Over time, the cactus islands increase in size, and the probability of a moth discovering them increases because the probability of encountering them increases. However, because moth abundance remains relatively low across the landscape, small scattered cactus populations persist because of the time lag between their recovery and growth, and the eventual discovery by the moths.

Two general rules to remember about predators and prey are these: predator abundance is always lower than prey abundance,[d] and predation almost never results in the elimination of the prey. A prey population will always grow in size before it becomes discovered or noticed or worthwhile to a predator. The prey will always outnumber the predator for that reason, but also because predators are almost always larger in body size than their prey and are therefore fewer in number. This strong relationship exists across all ecosystems and is rooted in very basic ecological principles concerning energy transfer across trophic levels in the food web. This abundance relationship extends to other population-level factors concerning the rate of reproduction of large and small organisms and their respective abilities to adjust to stress in the environment (discussed in chapter 1).

Predators act as controls on the population size and growth of their prey species, but they do not completely eliminate the prey species (except under very unusual circumstances). This is true whether the predator is a specialist on a single prey species or a generalist for many prey species. If the predator is a specialist, its success will reduce the abundance of the prey, which then leads to a decline in predator abundance. Once the abundance of the predator goes down, the prey-species population will recover, which will subsequently stimulate growth of the predator population. This cyclic behavior is based on the fact that predator abundance is always a response to prey abundance; that is, the growth of the predator population lags behind

---

c. The principles governing this process were formally outlined in: R. A. MacArthur and E. O. Wilson, *The Theory of Island Biogeography* (Princeton, NJ: Princeton University Press, 1967).

d. This does not apply to host–parasite relationships.

that of the prey species. When the predator is a generalist, the abundance of the particular prey species will be reduced to the point that encountering individuals of the prey species becomes uncommon. Individual predators will begin to focus on other sources of food. That is to say, they will switch their prey preference to a more abundant species with a higher encounter rate and thus higher predation success. Predation pressure will be reduced in the rarer species, and it will not be eliminated. Whether it recovers and becomes abundant once again depends on the abundance of the predator and its prey preference. If the predator does not readily switch back, abundance of that prey species can recover.

The act of intensive farming effectively modifies, simplifies, or destroys natural ecosystem structure by eliminating a wide variety of naturally occurring plants and animals and replacing them with a few species of domesticated crop species. However, early farm fields retained some attributes of natural ecosystems because of their small size and proximity to wild land. While farming is characterized by the production of *monocultures* of specific crops and those monocultures attract predatory pests (such as flower beetles, leaf caterpillars, sap-sucking insects, and root borers), the presence of large numbers of those pests also attracts their predators (such as spiders, wasps, other predatory insects, and birds).[2] These predatory species either make their homes in or near the farm and venture into the fields in search of food, or they subsist for part of the year in the surrounding habitats and disperse into fields as their food base increases during each growing season.[3] (See box 2-1.)

Just as the crop attracts crop pests, the crop pests represent a concentrated food source which allows the predatory species to dramatically increase in number over the course of the growing season. Because farms were small, the species that fed on crop pests could persist locally in trees and habitats adjacent to the fields and could move easily and quickly into the farmed areas to hunt.[4] A balance of sorts could be maintained because the predatory species did not depend on the farmed areas for the entirety of their subsistence and could forage for other prey species in nearby habitats. The farmer lost some proportion of the crop to pests but, with some precautions, the losses were tolerable most of the time.[5] Even today, farmers in many areas of the world have a good working knowledge of the positive contribution that predators play in reducing the negative effects of crop pests.

Early Uses of Technology

As farming moved into modern times, the costs and efforts associated with farming changed. Farmers bought seeds commercially for their annual sowing, they needed diesel fuel and gasoline for farm equipment, and they relied increasingly on human labor to exert more rapid manual control over common pests, especially weeds, but also insects such as caterpillars and beetles. The occasional outbreaks of insects or other pest species were managed with focused effort by humans to control the problem by manual removal of easily seen pests, precise application of general-purpose, naturally occurring toxins such as sulfur, and the physical pulling or hoeing of weeds. The objective was control and not eradication, which was an obvious impossibility with the limited technology. A significant part of a farmer's costs were in paying for human resources both to produce the crop and to harvest it. Farming remained a relatively small-scale enterprise, although farmers increasingly were producing goods for commercial sale and not just for their own families.

The farmer of perennial crops, such as grapevines and fruit trees, spent the winter trimming and preparing the dormant plants and fields for the coming season, and this also required manual labor. With long-lived perennial crops, the growing season was somewhat simpler in many ways, with thinning and harvesting the main production activities. Weed control and careful observation for pest outbreaks were regular chores.

The farming of annual crops (such as grains, potatoes, and tomatoes) always involved the production of herbaceous and fast-growing plants that produce high-quality food products in a short amount of time. These foods are very attractive to insects and other pests because of the high nutrient content that makes them so valuable to humans. If a particular pest became unacceptably abundant and destructive, the farmers could consider other control options, such as rotating crops from one year to the next. By annually alternating with a second crop that was unpalatable to pests that specialized on the first crop, those pests would disappear or be reduced locally to very low numbers after a year without a concentrated food base. This farming practice was not available to the farmer of perennial tree crops, who had to devise other ways of managing persistent pest problems.

As farm technology improved and the farming effort moved inexorably to larger and larger scales, the sizes of monoculture fields became larger as well. Commercial farming was most profitable if large acreages could be devoted to a single crop rather than many small plots of many different species. That is, increasingly, commercial profit lay in specialization. Unfortunately, the cultivation of a large expanse of a single crop, rather than a mosaic of many different crops, inevitably changes the dynamics between the crop and the living components of the surrounding ecosystem. First, a monoculture creates a single uniform food source that attracts and encourages specific herbivores, and it creates a condition of almost unlimited food availability. Second, as the size of the field increases, the distance to the surrounding ecosystem increases and that represents the source of species that might prey on crop herbivores. The interaction between predators and their prey occurs in a predictable fashion: the predator must locate the prey population, and this can't happen until the prey population grows large enough to be detected. Also, the greater the distance to the prey population, the greater the time needed for a predator to detect it. Additionally, it's important to remember that prey *populations* are detected by *individual* predators, which means that the control of the prey population depends on the reproductive rate of the individual predators (i.e., how fast they are able to multiply).

Thus, as fields of particular crops increased in size and as ecosystems were pushed back at the margins, the greater the opportunity for increased problems with herbivores and the lower the probability that natural predators could find them and control them. The common and potentially manageable insect pests soon became much more numerous nuisances that could greatly reduce the production of a field unless farmers increased the intensity of their control efforts. The change in the spatial scale of farming was and is characteristic of modern commercial agriculture, and it marks the transition to a much greater need for the control of crop pests, especially insects.

Farming Now

In the twenty-first century, the majority of modern farmers in the United States prepare for the coming growing season very differently than in the past. At the end of the growing season in the fall, the soil is typically tilled with all crop biomass incorporated into the soil and the land is left completely bare and fallow over the winter. For some crops, the residual biomass might be burned, which can allow for some nutrients to be returned to the soil quickly, or the biomass might be left on the surface over the winter as what is known as "conservation tillage."[6] Ideally, no insects or rodents—indeed, no living things of any kind—are supported or encouraged in the fallow soil or fields over the winter months. In this sense, modern farmland is "dead" for many months out of the year. Only during the growing season is it alive and then only to produce a single desired crop.

The practice of fallow farming is an attempt to interrupt life cycles of pest species by eliminating the possibility of overwintering on crop residue and to facilitate the decomposition of crop biomass and the return of nutrients to the soil. While farmers go to great lengths to attempt total control over the crop pests, a small portion of those pests do survive long enough to lay eggs, or they remain dormant in the soil or on the stems of the plants over the winter months. Thus, one of the farmer's fallow-season objectives in pest control is to eliminate hiding places, and the most obvious approach to accomplishing that is to eliminate all biological material. Even though many pests can overwinter underground, fallow farming contributes to control efforts because the application of preseason chemicals is facilitated when those chemicals can be put in direct contact with the soil where the pests are hiding. It is important to note that the "health" of the soil—the abundance and diversity of the microbial and invertebrate community—is of secondary importance to the more immediate need to control crop pests.

Modern technology has provided today's farmer with a new arsenal of chemical weapons that can be used to "prepare" the soil for spring sowing with attempts to control the residual pests that have survived the winter months.[7] Chemicals are applied, such as fumigants to control fungi and nematodes and pre-emergent controls to reduce the abundance of germinating weed seeds. For some crops, farmers might allow the weeds to emerge through the winter, then cultivate the soil in early spring to kill them, and then allow the next generation of weeds to emerge before applying a broad-spectrum herbicide before or, increasingly, after sowing the crop seeds. A growing number of farmers practice no-till farming to some degree because weed control can come after herbicide-resistant crops have been planted. However, even as this practice has grown in popularity because of the benefits to the soil and the reduced costs of operating machinery, the appearance of herbicide-resistant weeds is reducing the viability of no-till farming for some crops.

Once the spring crops begin to grow, some pesticides may be applied in anticipation of the emergence of known pests, particularly insects, mites, and other arthropods, while other chemicals are applied once a pest has been observed and identified. Some of the pesticides are relatively specific, such as preparations containing *Bacillus thuringiensis* (Bt) for killing herbivorous butterfly and moth larvae, and some are broadly toxic, such as the herbicides glyphosate and glufosinate, which kill most plants. The usual approach is to monitor the fields until certain species reach a threshold density or a certain growth stage, upon which pesticides are applied to the entire field. This may be accomplished with aerial application or tractor-mounted sprayers, but some chemicals can be included in the irrigation supply if roots are being treated or if the need is for the plants to take up the pesticide into the plant tissue.

Over the course of the growing season, depending on the crop, a farmer may make several applications of different chemicals, each for a different purpose, to control pests. In some areas, cotton has been reported to receive up to 30 chemical applications over the course of the growing season, but 12–15 is probably more common.[8] If the farmer did not apply chemicals, the productivity of the field would be so low that harvesting the crop would be pointless. In the case of fruit, not only would the usable crop yield be a very low percentage of the total, but the harvest would contain damaged fruit and fruit with insects, scales, mites, worms, and other unsavory living things that would require additional cost to remove before what remained of the crop went to market.

Pesticides Are Not Antibiotics for the Farm

It is important to note that the pest-control chemicals used in agriculture are fundamentally different from, for example, medicines used by humans. In the pharmaceutical world, each newly developed drug is patented and sold at a premium price to generate profit for the company, and also to recoup the investment costs of research, development, testing, permitting, and marketing that went into the production of the drug, which often exceed $200 million. However, after 12 years the drug patent expires (in the United States), and generic versions can become readily available at greatly reduced prices. While most drugs continue to work indefinitely, those that combat bacteria and viruses may lose effectiveness as pathogens become resistant over time. Even this resistance takes many years to develop unless the drug is used widely and intensively, causing resistance to develop more quickly.

The long-term use of pesticides in the world of agriculture is in stark contrast to that of human medicine. Most pesticides have a very short lifespan—five years is not uncommon—and farmers see many of the chemicals in common use become less and less effective over a short period of time.[9] This means, first, that new chemicals must be added regularly to the agricultural arsenal and, second, that farmers rarely enjoy the luxury of buying *useful* generic chemicals at low prices (the herbicide glyphosate is a rare exception). The pests they battle quickly become resistant, and the old weapons of war become increasingly ineffective and the new ones increasingly expensive. The middle ground of inexpensive and effective pesticides is increasingly short-lived.

Agrochemical producers have well-funded research and development centers for creating new, more-effective pesticides, but the short-term outcomes are always the same: every new chemical is expensive and farmers have little recourse but to use them or lose their crops to the emerging armies of resistant pests. In an ever more expensive world of chemicals, federal price supports and subsidies moderate the increasing cost of producing food and fiber in the United States. In 2007, US domestic pesticide sales were $12 billion (up 11.5 percent from 2000) and accounted for more than 32 percent of all sales worldwide.[10] The United States uses almost 40 percent of all herbicides and insecticides sold in the world, with the agricultural sector using almost two-thirds of the pesticides sold in the country. This continuously increasing and disproportionate demand for pesticides is due to three main factors. First, the United States is the largest producer and exporter of agricultural goods in the world and therefore consumes more resources in the production of those goods. Second, agriculture yield per hectare in the United States is greater than that of other large countries. Arable farmland under cultivation in the United States has declined from 0.62 to 0.52 hectares (about 1.25 ac) per person from 2001 to 2011,[11] while total productivity increased by 7.7–9.2 percent from 2000 to 2010.[12] In comparison, agricultural production in the Russian Federation (0.86 ha per person) and China (0.90 ha per person) has held steady.

Third, the United States is more dependent than other countries on synthetic pesticides for agricultural pest control. This may be an indirect result of being one of the centers of development and deployment of agricultural technology, but it may also be the direct result of the decades-old drive to increase production levels on arable land. This combination of changes to farming has created conditions that favor rapid evolution of pest resistance that is unparalleled in magnitude anywhere else in the world. There is no doubt that other regions of high agricultural production also use large quantities of pesticides and are experiencing pest resistance problems (e.g., cotton in Uzbekistan), but not on the same scale and level of diversity as in the United States.

The Changing Landscape of Farming

Farming statistics for the United States can be alarming. The marketing image of the farmer—a traditional, hard-working nuclear family—is still accurate for the majority of farmers: there are about 2.2 million farming families in the United States today. However, the reality of the modern farming family is not quite as glamorous as that depicted on TV shows and advertisements. About 60 percent of American farms have a "gross cash farm income" (GCFI) under $10,000, and 91 percent of all farms have a GCFI of less than $250,000.[13] Many if not most of these farmers do not farm as their primary economic activity. These small farms, whether private or commercial, are responsible for only 23 percent of the agricultural production in the United States.[14] Therefore, the vast majority of farms are small and are responsible for less than one quarter of total farm production. Also, 54 percent of all farms are in the category of "small non-commercial farms" and they produce only 1 percent of that 23 percent total production value.

From these statistics for the United States, we see that farm production is skewed toward large and very large farms, which make up only 9 percent of the farms but produce 77 percent of the total production value. (In fact, the largest farms make up 2 percent of the total, but produce 47 percent of the total production.[15]) This small proportion of farm enterprises accomplishes this feat using less than 50 percent of the arable farmland. Obviously, large commercial agriculture is highly intensive, production-oriented farming that is capable of extracting very high yields per acre. This approach toward farming, a wholesale change in the scale and intensity of operations, has been touted as a more efficient method for producing the agricultural products needed for a growing national and global population. For example, large commercial farming operations (that is, those that are much larger than one family could possibly manage) can make economic decisions that would be impossible for a family operation. Buying million-dollar pieces of equipment, building housing for multiple work crews, drilling new wells, and making marginal profits per acre on huge expanses of land are all activities well beyond the scope of the traditional farming family.

The shift from small family farms to large commercial farms has been a steady process for several generations in the United States, but it has greatly accelerated in the past 30 years due to economic policies that favor large-scale operations.[16] In the 35 years from 1982 to 2007, large farming operations (more than $1 million GCFI) more than tripled in number (to about 50,000), while the market value from small farm production fell from 42 percent to 23 percent of the total.[17] Thus, the perception that the US domestic food and fiber supply is the result of the hard work of small independent farmers making a living off the land and providing for the rest of us is increasingly anachronistic.

In terms of the production of important commodity crops, the "modern farmer" is a corporation that does not produce food for a single family or even for a village. The modern farm and its equipment are specialized for the production of only one or just a few crops. The process of farming is highly mechanized and geared toward the sowing, cultivating, and harvesting of vast expanses of farmland, expanses well beyond the grandest visions of the traditional farmer or the financial capacity of the modern family farmer. The process has been simplified such that each item of the equipment itself often has only a single purpose, and the size of the equipment is commensurate with the scale of the farming effort. One consequence of this change in scale has been a change in attitude toward the land itself.

The traditional family farmer of a century ago depended on the land for sustenance; the soil provided the food that the family ate. When human population density was relatively low, the process of farming was not overly destructive to the soil. This was true for several reasons; mainly, the family farmer rarely had the capacity to damage the soil on a large scale. The localized farming effort was restricted in scale and any damage to the soil was a small patch in a large sea of undamaged soil. If the soil were damaged by overuse, the farmer could move operations a short distance while the damaged soil recovered naturally. This is not to imply that the cumulative impact of family farming practices was never destructive nor ill-informed (the Dust Bowl era is a testament to that), but the need and obligation to protect the soil for future generations was a central tenet in the farming ethos.

Fast forward to the modern era in which multiple and massive engines of farming rumble efficiently across the fields, ripping, plowing, sowing, spraying, and turning the soil. The entire sea of soil is under cultivation; there are no places on the modern commercial farm that are exempt from the disturbance. Soil is not allowed to recover after it has been depleted of its nutrients and damaged from the plowing and mechanical activity. Every acre is cultivated and every pound of production is wrested from it, removed, and sold in the markets far from the place of production. Some parcels are allowed to "rest" for a season, but recovery for soil is not simply a matter of lying fallow.[18]

In a typical annual-rotation field (such as one used for corn or tomatoes), the soil is cultivated up to the very edge of the field, which is typically a road. No weeds, shrubs, or trees are allowed to exist between the road and the field. Where buffer zones are planted between fields or along roadside verges, nonnative species of trees or grass are used to reduce wind or water erosion. Ideally, from the production perspective, not one square meter of usable soil should be uncultivated. This "clean farming" practice maximizes production and is, simultaneously, an effort to keep weeds at bay by eliminating patches adjacent to fields.

Another goal of modern commercial farming is to make use of every portion of the plant whenever possible. For example, in the past, only the ear of corn was removed from the cornstalk. The rest of the plant remained in the field. Often, pigs and chickens were allowed to forage in the fields to glean whatever corn they could find, and the cornstalks were allowed to decompose. Between the pig and chicken manure and the decomposing stalks, important nutrients were returned to the soil and a large number of invertebrates and microorganisms thrived on the decomposing biomass. Modern agriculture takes a different view, and for very practical reasons. First, the agriculture of pig and chicken production has also shifted to specialized commercial farms, and the animals are no longer allowed to forage freely, much less in the cornfields. Second, it is increasingly common for the cornstalks also to be harvested and used for other purposes such as forage (for livestock production), biofuel, and fiber.[19] All of the soil nutrients that were taken up into the corn plant during the growing season are removed from the field at harvest, and this rapidly depletes the soil of nutrients for the next season, which necessitates the addition of fertilizers to the fields. These fertilizers might be manure from livestock facilities but are more likely to be synthetic, commercially produced fertilizers. In short, modern farming practices tend to be damaging to the quality, structure, and health of the soil (see chapters 13 and 14).

Growing Food or Making Food?

The large-scale farm operation of today takes a reductionist view of agriculture: a farm is a food-producing machine, and this view is a direct outgrowth of the changes in farming scale and resultant changes to farming practice. The change in the philosophy of farming places the focus on the product and not on the process of production. While the Department of Agriculture, the Soil Conservation Service, the cooperative extension agent, and the small farmer would insist otherwise, the soil is often viewed as no more than a nonliving substance, a matrix that acts as an anchor for the plants and from which the plant can extract the water and nutrients that are applied during the farming process. This hugely simplified view treats the land as no more than the context for growing food and the process of producing food as being very linear: spray and disturb the soil to eliminate weeds, add seeds, add water, add sunlight, harvest food, repeat *ad infinitum*. As a preferred practice, the environment aboveground can and should be sterile as food production is maximized with no interference from nature. This attempt at sterility is achieved with modern synthetic pesticides.

A healthy soil is anything but inert. A fantastic quantity and diversity of fungi, bacteria, worms, microbes, and roots inhabit healthy soil, creating a complex belowground ecosystem of herbivores, predators, detritivores, and decomposers that incorporate the dead plant material from above into the soil below.[20] (See box 13-2.) Plant nutrients are not only recycled back into the soil, but additional nutrients are made available by the multiplicity of activities of the soil's inhabitants. However, this community is easily disrupted by mechanical disturbance, compaction, flood irrigation, and the large number of aggressive and broad-scale chemicals used in modern agriculture.[21] The perception of soil as dead material is often not far from accurate, given the many insults endured by soil in the course of modern crop production. As we shall see, the attitude that the soil is largely irrelevant and that whatever the crop plant needs for growth can be supplied by the farmer has contributed to the modern dilemma of agricultural pests, especially chemically resistant pests. These are not unrelated phenomena.

In his book *In Defense of Food*, Michael Pollan has described the modern view of food as a list of nutrients packaged into a marketable unit.[22] In his opinion, we make choices about our food based not on quality or personal preference, but on marketability and consumer manipulation. Similarly, modern agriculture operates as if food production were a recipe—a minimal number of steps that require no real understanding of nature, of history, or, sadly, of long-term consequences. The soil is divorced from the plant, and the plant from the fruit, and evolutionary biology from the entire process. Any obstacle that reduces productivity can be overcome by identifying the obstacle and then creating a tool to eliminate it, though often without recognizing the nature of the obstacle and how interconnected it might be to the agro-ecosystem. This simplified approach to food and fiber production has had a dramatic effect on the worldview of the "modern farmer."

In the United States, if we consider the modern farmer as the entity responsible for the production of the majority of the food and fiber supply, then we must look at the very large commercial farming operation mentioned earlier.[23] There are more than 55,000 very large commercial farming enterprises (those earning more than $1 million GCFI) which represent about 2.5 percent of all farms. However, they produce about 45 percent of the total value of agricultural production on about 15 percent of the arable land. The very large farms tend to focus on commodity crops, such as dairy, beef, grain, and soybeans, and they produce these commodities in very large quantities.

In contrast, the smaller and noncommercial farms, representing 53 percent of the farmland, are far more likely to be involved in land conservation programs and account for 82 percent of the acreage in land-retirement and soil-conservation efforts.[24] Thus, on average, it appears that there is a distinct difference in attitude toward the inherent value of soil as farming efforts shift from the small-scale family farm to the large-scale commercial farm.

~

The farmer of 2014 faces the same basic problems as the farmer of 1944, but the problems differ in scope, magnitude, diversity, and complexity. The traditional farmer used lower-tech, lower-input approaches to farming and lost perhaps 32 percent of the crops to pests.[25] The farmer of today uses incredibly advanced and expensive technology to achieve higher yields, yet still loses about 26–40 percent (an average of 32 percent) of the world production of major commodity crops.[26] Thus, crop losses have not appreciably changed in 60 years, despite the development and application of massive quantities of the very powerful synthetic pesticides on which most farmers are now completely dependent. While the incredible yields that American farm operations wrest from the land are certainly impressive, one has to ask whether pest resistance, chemical pollution, and damage to the soil have been an acceptable price to pay for this productivity.[27] Indeed, the broader question would appear to be whether the inherent costs of this approach to agriculture can be sustained by the US economy (and the world's economy) for another 60 years.

Chapter 3

Survival of the Fittest: Darwin's Principles

The annual global sales for the agro-chemical industry now approach $40 billion.[1] How did it come to this? Three generations ago, very few chemicals were used to combat farm pests, and those were mostly natural compounds such as sulfur, or they contained elemental toxins such as copper, arsenic, and mercury, or they were some kind of plant extract. Since the late 1940s, with the introduction of synthetic fungicides and organophosphate insecticides, agriculture has become increasingly dependent on the use of chemicals to control the ever-increasing number of pests on the modern farm. There are very few farmers who are not financially dependent on synthetic compounds created by a chemical industry whose primary purpose—now—is to develop new chemicals or technological fixes to replace the old chemicals. The new chemicals will be used to combat the existing armies of pests, and the new compounds will be, even in the short term, largely ineffective. And there is no possible way to anticipate future farm pests except to know that there will be more of them and that the array of species will include those we already have. That is to say, even with the chemical technology of today, we will never eliminate the pests we already have and, in fact, they will only become worse.

To understand why this is true, we have to review some evolutionary biology concepts and the work of Charles Darwin. Darwin made two very astute observations about populations of species. First, as he traveled the world and collected a tremendous diversity of plants and animals, he came to the realization that a species is not "fixed" as had been commonly believed. That is, every individual in a population or species is, in fact, an individual and is not identical (with some exceptions) to any other individual in the population.[2] There is a wide range of sizes and shapes and differences among all the members of any and every population. As obvious as we might think this is today, most people in the Western world of Darwin's day believed in the "fixity of species"—that all things created by God were created according to a divine image, and one individual animal within a species did not and could not differ in any meaningful way from another. This belief permeated the scientific world as well.

Second, through his reading, particularly the writings of the British scholar and mathematician Thomas Malthus, Darwin realized that all organisms, large and small, are capable of exponential population growth. That is, as a population grows larger, it also grows faster, and potentially can grow to tremendous numbers if the resources it requires for survival are unlimited.[3] But resources are always limited in some way, and populations never realize that kind of growth potential for extended periods.[4] In fact, from one year to the next, most populations remain more or less the same size. While they can fluctuate in size annually or cyclically over longer periods, they typically have a stable point around which any fluctuation occurs. This fluctuation is related to changes in resource availability, year-to-year and seasonal variation, and changes in predator abundance. Thus, although populations are capable of increasing dramatically, they rarely do and then only for relatively brief periods.

These two fundamental observations, that individuals vary within a population and that populations can, theoretically, grow explosively (but don't), led Charles Darwin to three very important conclusions. First, if every individual in a population reproduces to its potential, yet the population does not grow, then the vast majority of the individuals must not be surviving or reproducing. A "vast majority" in this case means more than 99 percent of the population. For example, if a pair of frogs produced 50,000 eggs and even 10 of those young survived (99.98 percent mortality), the resulting 5 pairs of frogs could produce 250,000 eggs the next year—that is, five times more eggs. If survival in that year were the same, the 25 pairs of frogs could produce 1.25 *billion* eggs in the third year. Clearly, that pattern cannot be supported for very long or we would live in a world of frogs. Therefore, for frogs as for most species, particularly those that produce large numbers of offspring, very few offspring survive to the point where they reproduce. And this is very important: many offspring of most organisms *do* survive, forming a link in the food chain, but very few of them survive long enough to reproduce. In fact, *on average*, only two offspring from each pair of frogs survives long enough to produce the next generation. And thus frog populations, like most populations, remain relatively stable in size.

Darwin's second conclusion was that, if only a very few individuals of a species survive long enough to achieve reproductive success, there must be some underlying reason why those few and not all of the others survived. The reason cannot be just "luck" or a chance event; it has to be something systematic, something related to their particular set of skills or traits or parentage. Those few survivors must have, *on average*, a greater ability to avoid predators or to gain the food resources they need, or to find a mate or to defend a territory or, in general, to have a greater ability to withstand the stresses of the environment. At any time in any particular environment, certain characteristics will be more valuable for survival than others, so those individuals that survive do so because they possess the necessary characteristics for survival in that particular environment.

Lastly, Darwin concluded that, if an individual survives the rigors of the environment, the traits that enable it to survive are only useful for increasing evolutionary fitness if they can be passed on to

Box 3-1: Exponential growth

The concept of exponential growth was a revelation to Charles Darwin, and his appreciation of its power in nature formed the basis for the concept of natural selection. The calculations for exponential growth are no more difficult than calculating interest in a bank account, which provides a good analogy with nature. College students are often advised to begin to plan for retirement—advice which typically falls on unresponsive ears. But consider two accounts: in the first, $100 is added every month for 50 years for a total of $60,000 and, in the second, $60,000 is deposited once and no more is added. Given the same 5 percent interest rate (compounded monthly), which is the wiser choice? At the end of 50 years, account no. 1 will contain $266,865 and account no. 2 will contain $727,163. The reason is that the 5 percent monthly increment grows very slowly in account no. 1, but is at its maximum in account no. 2 for the entire 50 years. The greater the difference in size at the start, the greater the difference between the two accounts in the end.

In nature, this scenario has tremendous implications. Consider a population of two mice that can produce four offspring in a month and the four offspring (two pairs) can produce more offspring within a month. Assuming the mice produce offspring only once, the result is 4,096 mice in 12 months. However, if we begin with a population of 4,096 mice, after another 12 months the total is 16,777,216 mice. Each starting population had the same growth rate (doubling every month); the difference lay in the size of the starting population.

Darwin realized that it did not matter what organism was being considered, what the growth rate was, or what the time period might be; the pattern of growth was exactly the same, whether for ants, plants, or elephants. Importantly, of course, he also realized that such population behavior was rarely seen except in short-lived bursts of smaller organisms, and only under particular environmental conditions. Those conditions involve the removal of environmental constraints such as resource limitations, top-down control by predators, or abiotic controls such as drought or heat stress. In agricultural situations, all environmental constraints that can be alleviated have been alleviated for the benefit of the crop plants, which then grow to their maximum potential. Thus, it should come as no surprise that all organisms that *eat* crop plants will experience exponential growth as they encounter essentially unlimited resources on the farm. The rate of growth of a particular organism will be dependent on its time to maturity and its reproductive potential, so populations of smaller organisms with short generation times will appear to grow explosively.

the next generation. Thus, successful reproduction must be linked to the transmission of the valuable traits to the offspring. When he was developing the concept of natural selection, Darwin had no knowledge of genetics and the chromosome theory of inheritance (which was even then being investigated by his contemporary, Gregor Mendel), but he realized that merely *learning* to survive could not be sufficient; for the offspring to benefit from the ability of their parents to survive, the useful ability must be inherited. This leads us back to the first observation: for the most part, the different traits that enable individuals to survive must reflect the variation within populations that Darwin noticed as he explored the natural world.

These five postulates are the basis for the concept of natural selection and "survival of the fittest." All individuals of every species are subject to the stresses of the environment, and those that survive those stresses will have a much greater chance of passing along whatever particular trait they possess that enabled them to survive and reproduce. The subsequent generation will comprise individuals descended from those individuals best suited to their environment. Over time, the genetics of the population become fine-tuned to the stresses of that particular environment . . . provided the environment doesn't change too rapidly. (We'll come back to that point momentarily.)

These observations and conclusions make it clear that there are only two activities at which every individual of every species must succeed: feeding itself and surviving long enough to reproduce. The members of any particular generation are the descendants of those who were the best at obtaining food and reproducing successfully. For nearly all species, the primary day-to-day activity is the location and acquisition of food, in the case of animals (i.e., heterotrophs), or the production of food, in the case of plants (i.e., autotrophs). There is tremendous variation in the application of this theme, but the basics remain true for all individuals of all species. On the other hand, in order to survive in the longer term (i.e., passing one's genes on to the next generation), each individual must complete the process of producing an offspring generation. This process can be as simple as finding a mate and mating, and as complex as elaborate courtship rituals, fighting for and defending a territory, or building an attractive nest prior to the selection of a specific mate by the other sex.

The most complex physical structures and behaviors in both plants and animals can be understood by keeping this in mind. In plants, elaborate structures and mutualistic relationships exist to promote pollination and seed production, and the success of that endeavor is predicated on maximizing photosynthesis (i.e., producing food). In animals, the allocation of energy is devoted disproportionately to obtaining nutrients (eating) and to protecting reproductive investments. Beyond a doubt, the two greatest drives for all animals are eating and reproducing; the daily pattern of activity for literally every animal revolves around food and reproduction, usually in that order. As a corollary, the *rate* at which food can be acquired and the next generation produced can also be viewed as an adaptive trait for species survival. For example, in the event of a severe environmental stress that is lethal to the majority of the population, individuals that complete their life cycle, even if they die relatively prematurely, are "successful" if their alleles have been passed on to the next generation. Thus, organisms that mature rapidly or that can reproduce quickly are likely to adapt quickly to those environments characterized by severe or unpredictable stress events.

Darwin's concept of natural selection can be summarized as: *adaptations are responses to environmental stress.* Although each adaptation can be viewed as a response to a particular stress, over the course of hundreds of generations, the individuals of a population or species are fine-tuned to their particular environment in the sense that they possess a complex suite of traits, all of which contribute to their "adaptedness." There is one caveat, which is stated in every textbook on the subject: the individuals of this generation are the result of selection in the previous generation. That is to say, offspring possess traits that enabled their parents to survive in response to the stresses faced by their parents. As a consequence, if the environmental stresses of the previous generation are not the same as those of the current generation, the offspring also face selection pressure, and what was adaptive in the previous generation may not ensure success in this generation.

The genetic variation among individuals in the population provides the grist for the natural selection mill. In fact, if environmental conditions are highly variable from one year to the next, survival of the population is predicated on the amount of genetic variation available for natural selection. In contrast, if conditions remain very consistent from one generation to the next, the population will grow to the extent that resources allow, and succeeding generations will encounter stress that is related to competition for resources (food) and mates. In other words, Darwin's principles—that all populations are capable of exponential growth, and that most individuals die before maturity—illustrate that every individual of every population will face strong selection for survival, and that every species in existence has a long history of successful adaptation to stress. In fact, no organism in existence today has ever failed to adapt to the challenges of the environment!

In the context of the Red Queen, the race for survival (adaptation) is fundamentally imprinted into the genetic code of all organisms, from bacteria to bunnies to barracudas. The ability to compete in the race is described by Darwin's five tenets of natural selection, but it also creates and influences the foundational concepts of community ecology and ecosystem theory. Ecologists have always been interested in what happens when a stressor disrupts the normal functioning of a natural community, because such disruptions shed light on normal community structure and function. As humans interfere more and more with normal ecosystem functions, the intricacy and complexity inherent in natural systems become apparent. Unfortunately, our newfound knowledge of the effects of disturbance has not shed much light on how to undo such disruptions or how to repair disturbed communities. And this lack of knowledge comes to bear as we confront our growing need to control the chemically resistant superpests we are creating by ignoring the basic tenets of evolutionary biology.

~

If we keep clearly in the forefront of our minds that all organisms face a battery of stressors at all times in their environments, that the "best" individuals in a population are the most likely to survive environmental stress, that they possess traits enabling them to survive stress that can be passed on to their offspring, that feeding and reproduction are paramount activities for all organisms, and that these criteria for survival are not only unavoidable but desirable . . . then we are ready to move on to talk about farming and pests.

An understanding of the evolutionary principles embodied by the Red Queen Hypothesis should guide our use of chemicals around the world. This is true of insecticides and herbicides (that is, any biocide), but also of synthetic chemicals of any kind that are disseminated throughout the environment as a result of human activity. Synthetic chemicals are relatively new stressors in the environment, and the inevitable result of widespread use will be resistance in organisms able to adapt to them and the potential loss of organisms unable to adapt to them. As we increase the number of synthetic chemicals in the environment, the number of stressors faced by all organisms will thereby increase, and those organisms best able to cope with the new and increasingly complex aspects of the environment will prevail. A potential scenario emerges of highly adapted and adaptable small organisms and fewer and fewer large organisms. The consequences of the loss of larger (and often predatory) species to the problems of controlling the smaller (pest) species will be made clear in later chapters. However, the loss of any species, whether herbivore, pathogen, or predator, can potentially lead to unforeseen consequences in natural environments and in human-dominated systems.

In the following chapters, we will focus specifically on the development and deployment of synthetic chemicals in agriculture, which began in earnest after World War II and which have had profound effects on our world. Although no one chemical in and of itself threatens the planet, the systematic and widespread application and dissemination of synthetic chemicals for a tremendous number of purposes, especially agriculture, has emerged as a global threat to all ecosystems and nearly all forms of life.

The application of synthetic pesticides was a relatively innocent and well-intentioned endeavor to improve our enjoyment of the outdoors as well as our ability to produce foods, fibers, and other agricultural products. In farming, synthetic pesticides appeared to be the answer to a number of problems, some very serious and some merely perceived to be serious. Today, it is possible for us to look back and assess whether this chemical path is worth the costs—and whether it is even achieving the desired objectives. In the end, we have to evaluate not whether the development of synthetic pesticides was good or bad, but whether we can heed the Red Queen and incorporate a working understanding of evolutionary biology into the modern world of food and fiber production. The answer will determine whether or not we can create a sustainable future.

PART II

Ignoring the Red Queen

Our greatest problems result from the difference between how people think and how nature works.

—Gregory Bateson

Chapter 4

Reductionist Farming:
Losing Ecosystem Services

Farming has always been a battle against the natural elements and against the natural predators of the crops. It is, in essence, an effort to overcome the limitations of the environment, to maximize productivity, to minimize limitations, and to eliminate anything that gets in the way. Over time, farmers created and manufactured technologies to ensure that water was in adequate supply to avoid drought and heat stress and to ensure a successful result for the growing season. Taming the elements involved the relatively easy tasks of engineering access to water through means such as dams and canals, and choosing or developing crops that were best suited to particular climates.

Dealing with natural crop predators was a different problem. These were living enemies that didn't fight back; they just absorbed the best eradication efforts of humans and persisted. It is important to understand crop pests, whether insect, fungi, or bacteria, as our competitors and not as agricultural diseases. While many crop pests are diseases of the plant, the role of the pest in the agro-ecosystem is exactly the same as that of humans: both pests and humans wish to consume all or parts of the plants. So, while we may approach the control of a certain crop pest as we would treat a disease, our ultimate goal is to prevent another species from consuming our food. By maintaining or improving the health of the plant, we obtain more food, and that can be accomplished by eliminating anything that reduces the productivity of the plant and of the field, a productivity that we have worked very hard to maximize.

Historically, we have grudgingly accepted losses in this battle, but we have endeavored to minimize those losses. In the past, a human harvest of 60–70 percent of the crop was considered acceptable and a success. But over time, with changes in human population density and technology, those attitudes changed. What if the human harvest could be greater than 70 percent or 80 percent or 90 percent? Is there any reason to think that we could not eliminate pests altogether and claim 100 percent of the productivity of the field? And as productivity increased with crop breeding, fertilizers, and intensification, the stakes became higher and what were once acceptable losses were no longer acceptable. Today, crop yields are so high and so valuable that *any* losses can be viewed as potentially unacceptable.

This can be easily demonstrated with maize (corn) production in the United States. Since 1940, yields have increased by 400–500 percent[1] and by 300 percent since 1960, and the numbers are continuing to rise.[2] Farmers in 1940 produced about 30 bushels/acre, but production topped 160 bushels/acre (about 4 tons) in 2012. Therefore, the loss of 20 percent of the total yield in 2012 represents the equivalent loss of 100 percent of the productivity of 60 years ago. That is, 100 tons of corn production in 1940 would yield 80 tons if the loss to crop pests was 20 percent. In 2012, the same acreage of corn might produce 500 tons. A 20 percent loss to pests would be 100 tons, but with 400 tons still making it to market. Thus, although yield would be five times higher, the loss of 20 percent (the entire production of the 1940 acreage) can represent an unacceptable economic cost to the farmer. For nearly any small farming operation, the lost income would represent the difference between making a profit or taking a loss for the year. With commodity prices for maize hovering around $300 per ton, more than six times higher than in 1960,[3] one can easily see how the loss of 100 tons of production would be intolerable.

So can we eliminate all crop pests and claim 100 percent of the productivity of our fields? The short answer is no, but it is important to understand the many reasons why this is true. Underlying that negative answer is the concept of *simplification*: of the ecosystem, of the crop, of our approach toward agriculture. To appreciate how agro-ecosystems function, we first need to review some basic characteristics of natural ecosystems as a basis for comparison. Natural ecosystems are incredibly complex, and their productivity is a function of that complexity. The interconnectedness of the activity patterns of species (i.e., the food web) involves large numbers of both negative and positive interactions. Negative interactions between species (e.g., predators eating prey) are sources of environmental stress and are very likely to be balanced by positive interactions that alleviate that stress. In many ecosystems, the number of positive interactions might well outnumber the negative interactions.[4] Positive interactions are best known as *mutualisms*, such as bees pollinating flowers or the mycorrhizal fungi on roots providing inorganic molecules for plants, and *facilitation*, such as nurse plants creating less stressful habitat patches that favor the germination and growth of other species. In a natural ecosystem, the number of species, their abundance, their functions in the system, and the strength of their interactions may have evolved over millions of years. Removing one or more of the species would set into motion a complex set of adjustments among all species related to the use of resources and the abundance of each species. These adjustments might take years.

In contrast, agro-ecosystems are incredibly simplified systems with one producer (the crop) and, ideally, no herbivores and therefore no need for carnivores.[5] In fact, the farming ideal is soil and crop and no real ecosystem at all. But that ideal is "false to facts,"[6] because plants cannot be grown without some interaction with the environment beyond water and nutrients. At the very least, soil and its components are necessary. For example, an estimated 92 percent of all plants show at least occasional associations with mycorrhizal fungi in the soil, of which there may be 10,000 species in the world. As just one example among many, wheat grown with arbuscular mycorrhizae (fungi that penetrate the roots and provide increased water and nutrients to the plant) showed increased yield and resistance to drought stress.[7] Thus, a soil healthy enough to provide fungal associations will support higher levels of crop productivity, and those soils with sufficient complexity to provide positive interactions will be better soils for agriculture.

In ecology, the term *complexity* refers to the number of species in a system, in terms both of their interactions (e.g., the number of links in the food web) and of the amount of functional redundancy they represent (e.g., the number of species of insect-eating birds). Complex systems tend to be relatively resistant and resilient—two properties that are studied intensively by ecosystem ecologists. *Resistance* is the ability to withstand stress: in other words, how great a hit an ecosystem can take without being disrupted or unable to recover. *Resilience* is a measure of how quickly an ecosystem can return to some semblance of its previous state after experiencing a disturbance. Both qualities have been closely linked to the complexity of ecosystems because more-complex systems can (usually) absorb disturbance more easily without being disrupted.

Ecosystem complexity is often measured in terms of species diversity and functional diversity. High species diversity is not just how many species are present, but how evenly distributed they are. For example, a community with 10 more or less equally abundant plant species is more diverse than a community with 10 species in which one of them represents 90 percent of all the individual plants. High functional diversity means not only are all the different types of organisms present, but there are multiple examples of those different kinds of species. For example, there are over 250 species of native ants in California.[8] When diversity is high, the loss of one or more similar species is unlikely to create damaging imbalances in the community, but when diversity is low, the loss of certain species can have a disproportional effect because the functions they served in the community may be completely lost. Thus, resistance and resilience are measures of the *stability* of a system in the sense that such systems are better able to persist after stressful events, and this stability is closely linked to species diversity.

Even when ecosystems are diverse and undisturbed, some species can be more important to the stability of the system than others, and their loss can have dramatic effects. The concept of the *keystone species* captures this relationship; it was first described by Robert Paine of the University of Washington in his studies of the effects of starfish on coastal communities in the Pacific Ocean.[9] These marine communities are very diverse, with many species of algae, mollusks, crustaceans, and small fish. The starfish are predators mainly of stationary species such as mussels (mollusks) and barnacles (crustaceans). When starfish were excluded from certain areas of the habitat, those areas became overrun with stationary species, particularly mussels, which crowded out nearly all other species. Without adequate space, the algae species disappeared, and along with them, the food and protective cover for a large number of small organisms. The influence of the starfish was so important that it maintained the integrity of the community.

Without the predatory effect of the starfish, the community ceased to exist. This type of influence, of a predator suppressing the growth and abundance of its prey species, is termed "top-down" control because the influence comes from higher up the food chain. The effect of predation is to reduce the numbers in the prey populations, keeping them below the maximum sustainable level of the habitat. This can result in more available resources for other, less dominant species. In contrast, the abundance of herbivores and their predators is also determined by the productivity at the level of the producers (plants)—a form of "bottom-up" control. That is, climate and the fertility of the soil determine the quantity of plant biomass, which is the amount of food that is available to herbivores. We can see that the more productive an ecosystem is, the greater the abundance of herbivores and therefore the greater the abundance of carnivores. Those systems that are productive enough to support large numbers of carnivores should also exhibit some measure of top-down control by carnivores on the herbivores in the system. However, when the predators are missing, the controls over herbivores only occur as their populations increase in size, creating the bottom-up limitations of food and disease outbreaks. When top-down control of an herbivore is missing or removed, the population will grow unabated until some other limitation interferes with that growth. Typically, the next level of control takes one of two forms: either food becomes a limiting factor, resulting in high mortality and lower reproduction, or disease outbreaks become increasingly common because of the high densities of individuals, and thus mortality rates increase.

However, we must also remember that a habitat or community is part of a larger ecosystem, whether natural or agricultural, and is open to the surrounding environment. Movement is possible from peripheral areas into the system and from inside the system out to the peripheral areas. If population densities increase and food resources become limiting, individuals can leave in search of better habitat. Most natural areas are very heterogeneous, with resources patchily distributed, and this affects the distribution of each species within the region. Indeed, as seasons change and fluctuate from year to year, a species may go from locally abundant to locally extinct in the following year, but it may recolonize in the year after that. Different parts of a habitat may be favorable areas for a species, while other parts of the habitat might not. Ecosystems have been described as "habitat mosaics" because of this common and natural heterogeneity of resources that strongly affects the abundance of both plant and animal species over the course of time.[10] In such systems, the abundance of herbivores and the abundance of the predators that feed on them are both very dependent on the year-to-year variation in plant productivity, which in turn depends on rainfall, temperature, and other abiotic stresses. If herbivores do not find the necessary plant resources, the populations will decrease due to emigration, mortality, or low fertility, and the populations of predators will quickly follow suit. As the plant populations recover, following more suitable growing conditions, the herbivores will return. Critically, the herbivores will not return until *after* the plants have recovered (that is, there will be a time lag), and predators will not return until after the population of herbivores has recovered. This ebb and flow of producers-herbivores-carnivores has a cyclic effect on population densities. As herbivore populations increase in response to resource availability, their predators will increase after a time lag. As predators increase and herbivores begin to consume the resources, the population growth of the herbivores will slow and then decline. This will be followed by both a decline in the predator population and a recovery in plant abundance and biomass. When food becomes available and predation pressure has declined, the herbivore population will begin to recover again.

This cyclic behavior of predator and prey populations is rather common and is often very stable in the long term, but it often depends on the predators having very specific prey choices (see the *Opuntia-Cactoblastis* example from Australia, box 2-1). In systems where predators have a single prey species, the predator will decline in abundance *before* predation pressure eliminates all of the prey because the prey species becomes increasingly difficult to find and the predator does not have an alternative food source. When predators are generalists and are able to switch to other prey species if one becomes scarce, the predator and herbivore population numbers tend to be more stable in the short term.

Understanding these characteristics of natural ecosystems, their communities, and the species within them gives us an essential point of comparison with agro-ecosystems. While we do not normally think of farms in terms of resistance, resilience, stability, biodiversity, and the other ecological concepts discussed above, these simplified systems are governed by the same principles—and that has consequences for productivity and yields. In other words, farming attempts to create an agro-ecosystem that maintains a high-density monoculture, contains extremely low genetic variation, maximizes productivity with unlimited resource availability (i.e., bottom-up control), minimizes or even eliminates herbivory, functions without predation (i.e., no top-down control), yet operates with open

Box 4-1: Secondary compounds and crop plants

While plants are autotrophs and thus are capable of manufacturing their own energy supplies, they also form the base of the food chain, which means they are directly or indirectly food for all other organisms. And as organisms that cannot flee from predators, plants have only a few options for defense: physical defenses (e.g., thorns, prickles, and hairs) or internal and external chemical defenses. In general, plants produce two sorts of chemical compounds. Primary compounds are those chemicals that are necessary for day-to-day physiological functions such as photosynthesis, reproduction, transport of nutrients, DNA replication, and protein synthesis.[a] Secondary compounds are those chemicals that are not needed for daily function; these include alkaloids, terpenes, glycosides, and phenols. Not only do these compounds give the plants we eat much, if not all, of their flavor and character, but an estimated 50–70 percent of all new drugs are derived from secondary plant compounds.[b] These compounds are not produced for the benefit of humans, but for the benefit of the plant.

Secondary compounds are derived from primary compounds but have very different chemical activities and functions.[c] Alkaloids are derived from the modification of amino acids (building blocks of proteins). Glycosides are derived from sugar molecules. A significant amount of research has demonstrated the importance of secondary compounds as defense mechanisms against herbivory, particularly by insects.[d] Therefore, although atropine (a tropane alkaloid) has a strong effect on human cardiac function, it has a very different effect on insect metabolism. Caffeine (a xanthine alkaloid) is found in the leaves, fruits, and seeds of many plants and stimulates the human central nervous system, but it also acts to paralyze and kill insects. All of these compounds, including aromatic molecules and essential oils, are valued by humans for their culinary and medicinal uses, and all are very likely linked evolutionarily to the plant's defense against insect, bacterial, and fungal

a. G. S. Fraenkel, "The Raison d'Être of Secondary Plant Substances," *Science* 129 (1959): 1466–70.

b. D. J. Newman, G. M. Cragg, and K. M. Snader, "Natural Products as Sources of New Drugs over the Period 1981–2002," *Journal of Natural Products* 66 (2003): 1022–37.

c. Fraenkel.

d. P. D. Coley and J. A. Barone, "Herbivory and Plant Defenses in Tropical Forests," *Annual Review of Ecology and Systematics* 27 (1986): 305–35.

attacks. In fact, it is very likely that nearly every flavor and aroma in plants that humans find useful or desirable is related in some way to defense against herbivory. Other compounds may also be attractants for pollinators (such as floral scents) or attractants for predators of the insects eating the plants.

The process by which primary compounds are co-opted by the plant for other uses is relatively straightforward. If the plant has a mutation that results in a modification of a primary protein, thereby giving it a secondary function, and that function happens to decrease herbivory, then that plant will be more likely to survive and produce more seeds than plants that do not have that mutation. As with any mutation, if fitness is increased, the mutation is more likely to persist in the population and eventually become common. This modification of primary molecules is more common in plants than in animals because of the high level of genetic redundancy in plants as a result of *polyploidy*, chromosome duplication, which is found in about 80 percent of all plant species. Polyploid species are more likely to have duplicate copies of their chromosomes, and a mutation in one does not result in a loss of function, because of the remaining intact versions of the gene. Instead, if the mutation imparts additional function, as with plant defense compounds, then polyploidy becomes a favorable condition.

The plants that have been cultivated for human use have been highly modified to suit our taste preferences. Many wild versions of our present-day cultivars were somewhat sour, bitter, acrid, or fibrous, and certainly not as juicy, sweet, tender, or chewable as what we've grown accustomed to. Over time, farmers and breeders have chosen the more desirable traits in our food and fiber crops, especially plants that produce more of the harvestable product. However, many of those "undesirable" traits were a reflection of the defense compounds produced by the plants. As we have bred for plants with lower amounts of those secondary compounds, we have simultaneously bred plants that are less able to defend themselves from herbivory. Thus, modern breeds of crop plants are more likely to be attacked and, therefore, more in need of protection. For example, domestic varieties of rice have about 30 percent more carbohydrate calories than wild rice. Faster-growing plants do not produce as much lignin and are therefore easier to digest. All of the characteristics that make crop plants desirable for human consumption also make them very attractive to other herbivores.

borders to surrounding vegetation and habitats. Given what we know from decades of ecological research, such a system is, to put it generously, inherently unstable, and it will require tremendous external input (energy and resources) to prevent other characteristics of ecosystems from exerting influence on crop productivity.[11]

A reduction in species diversity necessarily results in a reduction in functional diversity to the point that very few of the normal stabilizing processes that occur in natural ecosystems are operating in agro-ecosystems. As aboveground diversity is minimized, so too is belowground diversity when the soil is plowed, turned, aerated, flooded, fertilized, and left without any plant cover for months at a time. The vast array of microorganisms in the soil is diminished, as are macro- and micro-nutrients, water-holding capacity, cation-exchange capacity, pH buffering ability, and resistance to erosion. Left behind are the organisms, such as fungi, bacteria, and nematodes, that can tolerate such extreme conditions and survive on the remaining nutrients. Similarly, the resulting highly simplified aboveground community is characterized by plant species tolerant of agronomic conditions (i.e., annual weeds that produce dormant seeds) and generalist herbivores (e.g., cutworms, grasshoppers, house mice). What are always missing from agricultural fields are resident populations of large predators (other than some raptorial birds), which can only persist in local patches of natural habitat.

In an agro-ecosystem, this loss of diversity is not necessarily undesirable. The goal of farming is production of a single crop, not maintenance of an ecosystem. However, achieving high crop yields in a simplified ecosystem becomes difficult because so many natural functions are lost.

A typical field begins in the early spring with a multi-species layer of young weeds. A broad-spectrum herbicide, such as glyphosate, is applied, or the field is lightly cultivated to turn the soil and prepare the field for planting, as with a blank slate. A crop may be sown as seeds or germinated in the greenhouse, as with field tomatoes, and then planted as small seedlings. Fertilizer is added before or after planting, and the soil moisture is kept high; that is, growing conditions are made to be ideal. There are no aboveground herbivores at this stage because the plant biomass is insufficient, or the field may not have attracted the adult insects, or the insects may not have made their spring emergence. Belowground herbivores may or may not be active at this stage, depending primarily on the soil temperature. As the plants grow larger and the biomass of the field increases, herbivores of different kinds will begin to appear, at first in small numbers, but they will rapidly multiply because of the lack of predators. These herbivore species will emerge from a dormant state in the soil or will migrate in from adjacent habitat. Their predators will be slow to arrive or will emerge only after the herbivore population is well established. However, a few predators, such as birds, may be naturally present if they are highly mobile.

In an ecosystem where they occur naturally, predators will respond rapidly to increasing populations of prey species. In his classic book *The One-Straw Revolution*, Masanobu Fukuoka described his techniques for rice farming in which natural predators were not eliminated as a consequence of pesticides or cultural practices.[12] For him, the presence of spiders, a highly effective predator in crops, was a sign of a healthy field. In combination with other natural farming methods, his rice production was as high or higher than on conventional farms that used pesticides and other techniques that simplified the farm ecosystem. He was able to control growing populations of crop pests because the growth of the predator populations was not far behind them.

Unfortunately, spiders are among the first casualties of pesticides. First, the spiders' prey are the targets of the pesticides and therefore their food source is greatly reduced. Second, spiders are arthropods and also vulnerable to many insecticides. Once the spider and other predator populations shrink, the top-down control of the prey is reduced or eliminated and the prey species can grow back unabated.

As a food source, such as the herbivorous prey species for spiders, begins to grow rapidly in response to an abundant resource base, such as a crop, the population growth response of the predators necessarily lags behind the prey. The predator population cannot begin to grow until its food source is already present and there has been ample time for reproduction. And the time lag will be greater if conditions hinder the predator from locating the prey: if the predator is not present and must migrate into the field after locating the burgeoning prey populations; if the distance the predator must travel to find the prey increases; if the predator is not highly mobile and does not disperse well; or if the population of predators in the region is low because agricultural expansion has converted natural areas to crop production. Thus, any farming activities that reduce resident populations of predators will slow the response of those predators and reduce the effectiveness of top-down control on crop pests.[13]

Modern agriculture is therefore an attempt to produce crops in ecosystems in which most of the natural controls have been eliminated and supplanted with a series of artificial controls. When pest outbreaks occur, the farmer cannot depend on natural controls because of the ever-increasing time lags between outbreaks and predator response. Historically, some measure of human intervention was always part of the process as farm workers managed weeds, removed diseased plants, and manually controlled isolated eruptions of particular pests until the natural processes could assist them. These techniques did not eliminate pests, but they were attempts to control their abundance and the subsequent crop losses. With larger fields and larger expanses of monocultures, it took farm workers longer to detect pests and so the costs in time and effort rose, as did the crop losses. Farmers needed faster techniques for pest control, and the most economical approach was to deal with the crop as a uniform system.

The development of synthetic pesticides after World War II changed the fundamental approach to farming. Pesticides promised a new era in which the farmer was not resigned to losing a portion of the crop to pests, in which the farmer was less dependent on costly manual labor, and in which profits would soar because of the increased yields being brought to market. With the emergence of new, powerful chemical weapons that appeared to eliminate crop pests completely, farmers were understandably quick to adopt this miracle of modern science. A farmer could now expect 100 percent recovery of the crop production and did not have to accept any losses. The new farming style attacked pests uniformly and indiscriminately by applying chemicals over the entire field, regardless of the population density of the pests. The field itself became a blank canvas that the farmer-as-painter could control completely, with little regard for the direct and indirect (and untested) consequences of the chemicals on the long-term health of the highly simplified ecosystem. After centuries of more or less firsthand personal contact with the soil and the crops, farmers were ceding much of the daily control of farming to chemicals. This represented a philosophical shift, but it arose as a consequence of the inevitable move from small-acreage family farms to large commercial enterprises. Synthetic pesticides, and chemicals in general, were seen as necessary to ensure reliable production of food and fiber on a scale that could never before have been imagined.

This approach to farming made sense; in the previous 50 years medical science had accomplished an extraordinarily similar feat in the battle against infectious diseases. From about 1900 to 1950, human intervention using modern chemistry had brought under control such dangerous and debilitating diseases as influenza, measles, polio, diphtheria, whooping cough, scarlet fever, typhoid, and tuberculosis.[14] The breakthroughs had come one after another in rapid succession and created the euphoric feeling that science was a conquering force that could end human suffering once problems had been identified and the tools of modern science could be applied. The successes of modern medicine could be seen as a template for a modern approach toward controlling the ills that plagued farmers. That is, crop pests are an infection and can be treated as one would treat a disease. Unfortunately, while we can view both the human body and an agricultural field as ecosystems, the comparison is far from exact.[15]

While humans live in populations, a disease-causing pathogen lives within an individual and creates a population within that individual. As the pathogen spreads through a human community in an epidemic, it forms a population of populations—that is, a collection of individual populations of pathogens that do not interact easily with each other because they are isolated in different individual humans. This is similar to the concept in ecology of the *meta-population*, which has been very useful for describing the structure and behavior of species that live in highly fragmented and dispersed habitats. In a meta-population, there are large and small populations with some movement between the units (depending on distance, population size, and a few other factors), but which, for the most part, behave independently. Disease in humans is very similar, although transmission of the pathogen from person to person is a necessary component for the survival of the pathogen.

When a medicine is given to a human, the chemical enters a (nearly) *closed system*. The chemical is applied to a single physiological unit and influences only the specific pathogen population within that unit.[16] The "disease" is caused by a pathogen that has overcome the defense mechanisms of the host, typically because the body cannot cope with the growth rate of the pathogen, and the chemical slows the growth rate by reducing the pathogen population, which allows the immune system of the body to regain control. The effects of the disease are mitigated; the pathogen population is reduced to very low levels and possibly eliminated; the system regains health and is not subject to further outbreaks of that disease, even when the application of the chemical is stopped, because of the antigen-antibody response of the immune system. By and large, the effect of the medicine is permanent, and reemergence of the disease will be a function of mutations in the pathogen and not merely of reinfection. All of this takes place independently of all other pathogen populations contained in all other human bodies.

Agricultural chemicals are applied to *open systems*. The application of a pesticide to a particular field is not an isolated event, nor is the field an isolated entity. There is constant movement of animals and plants from the field to the outer areas and from the outer areas into the field. Even the application of the chemical crosses the borders of the field and influences adjacent crops and organisms. The control of the target organism by using the pesticide can be immediate and impressive, but it is never complete or permanent.

The explosion of a pest population, like a pathogen causing a disease, is a direct response to a seemingly unlimited resource supply. The application of a pesticide greatly reduces the pest population, just as the medicine reduces the pathogen population. However, once the human system has overcome the pathogen, it is prepared for and protected from further infections by that particular pathogen. This is where the pesticide-as-medicine analogy falls apart. First, the natural immune system of the agro-ecosystem is lost because the predator species are typically greatly diminished, which would be analogous to a medicine reducing the number of white blood cells. Second, as soon as the pesticide becomes ineffective or is inactivated, which can literally be as soon as the sun comes up the next day, the next wave of invasion by the very same pest species can begin. The invasion will come from the neighboring untreated fields or from natural populations in adjacent areas or from survivors in the treated fields. The period of relief for the farmer will only last until the small initial population recovers sufficiently to have a negative effect on the plants' growth and production.

Of course, the long-standing solution to this recurring invasion scenario is to make serial applications of the chemical. Rather than paralleling the use of medicines to treat diseases, this situation is more akin to using antihistamines to treat an allergy. As long as the foreign protein is in the environment and continues to create an allergic response in a person, the antihistamine has to be applied to reduce the allergic reaction, but it does not "cure" the allergy. Similarly, if an agricultural field is open to the environment, it presents an untapped resource for any herbivore capable of exploiting it. As one pest population is treated, another will soon follow because the resource remains available. Agriculture creates massive, readily available resource bases for herbivores and, just as air will rush in to fill a vacuum, any organism that can take advantage of that resource will be evolutionarily rewarded. If the primary day-to-day activity of every organism is to obtain nutrition, then one would expect that all organisms capable of exploiting agricultural resources will be a persistent threat.

The far more troubling complication of the farm-as-an-open-system scenario is that both the populations of pests in the fields and those that migrate in after each chemical treatment possess varying degrees of genetic variation. Genetic variation fulfills the first tenet of natural selection. Even if the first, second, or tenth application of a chemical kills every individual in a pest population, eventually there will be one or more individuals that possess an allele of the gene that provides resistance to the toxic effect of the pesticide. This is the invariable result of chemical applications in agriculture. The progression of events that led to larger farms, increasing simplification of the agro-ecosystem, the use of very powerful toxins uniformly and repeatedly applied has created a scenario in which either every individual of the pest population must be eradicated every time or resistance becomes inevitable.

Chapter 5

A Weed by Any Other Name: Monocultures and Wild Species

The control of weeds on the farm is a major expense and the problem is continually growing. Where do these weeds come from and why are they so problematic on farms? As many as 3,000 species of nonnative plants have become established and common enough to be considered "naturalized" (i.e., permanent residents) in the United States.[1] That may seem like a large number, but 3,000 represents only 10 percent of the number of nonnative plants commercially available. It has been estimated that more than 1,500 (more than 50 percent) of all the naturalized species so far introduced to the United States came through "intentional" introductions of one sort or another. These introductions include food crops, herbs for medicines, herbs and spices for cooking, forage for livestock, and ornamental plants, as well as plants introduced for erosion control or even because they reminded human immigrants of their homelands. Many introductions might have been accidental or associated with agricultural activities: cows and sheep carry weed seeds in their fur and feces, and weed seeds are frequent contaminants in the crop seeds that we plant.

In California, the introduction of nonnative plants, whether deliberate or accidental, has been happening since before 1769, when the first Spanish missions were established. The mission buildings were constructed primarily of adobe bricks made from local clayey soils and held together with straw and other plant materials. At least three species of European weeds were already present in the California landscape at the time of the first mission; the plant remains were found in the earliest adobe bricks used.[2] It is possible that the first European weeds in the western states were introduced as early as 1535 by livestock that accompanied the Spanish explorer Hernán Cortés.

In the nineteenth century, many undeveloped areas of the United States, particularly in California and the Midwest, were being rapidly converted from wildlands to regional centers of agricultural production. The seeds for the crops being grown there were often produced elsewhere in the world. For example, to feed the growing numbers of livestock and draft animals in the United States, alfalfa was grown in many places across the United States. The alfalfa seed imported to the United States was produced mostly in Turkey and contained seeds of many weeds commonly found in the Middle East. Whether because the demand was so high or standards too low, over time the quantity of weed seeds in the crop seed rose to the point that farmers of alfalfa (and other crops) complained to state and federal officials. In some cases, there may have been more weed seed than crop seed in some shipments, and this was having an effect on the quality of alfalfa being produced. In the 1800s, the concern was not centered on the introduction of weeds, but on the reduced crop production and quality that is inevitable with heavy infestations of weeds.

The Seed Purity Laws enacted in the United States at the end of the nineteenth century and later codified in the Federal Seed Act of 1939 were an attempt to control the number of weed seeds contained in seed lots of crop species, whether imported or domestic. That is, when crop seeds are marketed the labeling must be accurate, and limits were established for how many weed seeds could be present for particular crop species. In the United States, seed lots may have a zero-tolerance criterion for noxious weeds, or a limit of two seeds per sample for prohibited species, or a required accurate labeling of the percentage content for restricted species.[3] In Australia, similar standards have been adopted for imported crop seeds with tolerances that range from zero to a maximum number of weed seeds per sample.[4]

To understand why weeds are so commonly associated with crops, we must consider the environment of a farm before, during, and after the growing season. The farmer's goal is to eliminate all competition for the resources the crop needs and this can only be achieved by removing all other plant species. Then, once the crop species is in place, the objective becomes to create an environment of unlimited nutrients, water, and sunlight to realize the maximum growth and reproductive potential of the crop. The farmer tries to minimize year-to-year variation in production, to stabilize yield such that it is very predictable. This process of reducing "environmental stochasticity" alleviates the unpredictable negative effects of the environment.[5] The resulting field conditions are the best possible not only for the crop but for almost *any plant* that can cast a seed into the soil.

When ecologists discover a nonnative plant species in natural or seminatural habitats, they assume that the newly introduced plants will require a period of adjustment to the new environment before they can spread and become invasive. A newly introduced species may or may not successfully adjust, or it may take many generations to adapt to the habitat. In contrast, weed species introduced in contaminated crop seed lots are almost certainly agricultural weeds in their country of origin and therefore do not have to adjust to the new agricultural conditions. Regardless of the crop, agriculture has similar objectives worldwide and so the growing conditions encountered by weeds are similar worldwide. Thus the weeds in a wheat field in China will find ideal growing conditions in a wheat field in Kansas, and vice versa.

Agricultural weeds are wild species that have adapted to human-dominated environments and are highly tolerant of the annual dis-

turbance cycles that characterize crop production. Nonetheless, although weeds are well adapted to agricultural conditions, they are
also very capable of continuing to adapt as conditions change. Over
the centuries, farmers have tried many creative ways to reduce the
negative effects of weeds, yet the weed species have successfully
adapted to those efforts.[6] When growing among crops, weedy annual plants have three strategies that allow them to avoid detection:
first, many species physically resemble the crop, as with barnyard
grass and rice plants; second, the seeds of many species resemble the
seeds of the crop plants, as with vetch and lentil seeds; and third,
many weed species produce seeds that fall to the ground and remain
dormant in the soil until the next growing season, as with jointed
goatgrass in wheat fields. One way or another, crop weeds often
resemble the crop they infest and avoid detection long enough to
produce viable seeds. For some crops, genetic variation for a particular physical trait has been introduced that allows farmers to dis-
tinguish crop plants from weed plants, but the weeds adapt quickly
to mimic the changes in the crops. This is easy to understand in
terms of natural selection: those individuals with mutations for the
new appearance will live longer and produce more seed than those
without. In subsequent seasons, more and more of the weeds will
take on the appearance of the crop.

The ability of weedy plant species to make rapid adjustments
to the conditions of the agricultural field is an adaptation in itself.
When the growth form or growth rate or developmental pattern
of a plant varies with the environmental conditions, the plant is said
to be *plastic*. That is, the appearance (phenotype) of the plant is
malleable (plastic) and changes in accordance with the conditions
or limitations of the location. Plants in conditions of resource stress
may develop faster, produce seeds earlier, and become dormant for
a longer part of the year, but they may grow and behave differently
under resource-rich conditions. One characteristic of plants that may
favor plasticity is *polyploidy*, that is, having more than two copies of
each chromosome.[7] The additional sets of alleles may provide for a

greater range of genetic expression in plants and a greater ability to survive and adapt to a variety of environments than those of diploid organisms with genetic coding for very few phenotypes and perhaps less ability to adjust or adapt to dramatic changes in the environment. (See box 1-2.)

In the diploid genome, expression of alleles is restricted to two or three phenotypes as a result of allelic dominance and co-dominance. For example, common garden flowers such as snapdragons or four o'clocks can be red, white, or pink, and that is the limit to the color variation. In contrast, polyploid plants commonly have four or more copies of all chromosomes and can possess many different alleles for a single gene. For example, a tetraploid plant species with four copies of all chromosomes has the potential for 16 different allele combinations. If these alleles can be expressed in different ways (i.e., "turned on" and "turned off"), the plants could produce seemingly adapted offspring in as little as a single generation. The potential for a single plant to express high intrinsic levels of genetic variability implies a potential for very rapid adjustments to changing conditions. Thus, high phenotypic plasticity can buffer the polyploid plant from natural selection pressure by allowing for rapid physiological adjustments from one generation to the next.

All flowering plants are likely to have experienced a genome-doubling event at some time in their evolutionary history.[8] This may be the result of either hybridization between two different species or a doubling (or quadrupling or more) of the genome as an accident during the production of ovules or pollen during reproduction. If such events happened far enough back in evolutionary history, previously identical chromosomes may not appear similar now, and the organisms may be classified as diploid. Polyploidy events result in the instantaneous formation of a new species and may be instrumental in the ability of those new species to survive extreme variations in resource availability and climate patterns. Although little research has been done in this regard, polyploidy should generate a greater range of phenotypic expression (i.e, phenotypic

Box 5-1: Adaptation, plasticity, and mutations

In the movie *Jurassic Park,* when mathematician Ian Malcolm comments that "Life . . . finds a way," he is making a reference to the undeniable power of organisms to adapt to the living conditions they encounter. Simply put, an *adaptation is an evolutionary solution that reduces an environmental stress,* and natural selection is the unquestioned mechanism behind adaptation under most circumstances. The process of adaptation is remarkably simple (it is outlined in chapter 3) and essentially requires nothing more than genetic variation in a large population. From that point, the mechanical process of differential reproductive success by the survivors results in a genetic shift in the population from one generation to the next as those individuals with the favored genotype make up more and more of the population.

All genetic variation in all populations is ultimately derived from mutations to the DNA molecules in individuals in the population. DNA encodes instructions for the cells to produce proteins, which are the primary functional molecules of the cell and body. Proteins have very precise structures and functions, and even a small change to the structure can alter the function. Most mutations to the DNA that result in structural changes also result in functional changes to the resulting proteins, though the change in functionality may be very slight. The odds that a random mutation will result in a more effective protein or a better-adapted individual can be compared to the odds of randomly selecting the right numbers for a winning lottery ticket: the odds are very nearly zero. But this is the difference between the statistical concepts of permutations and combinations. Essentially, we cannot know in advance which specific lottery ticket will win, but we know that, if enough tickets are sold, one of them will be the winning ticket. Given high ticket sales, the odds of *any specific person* winning remain very close to zero, but the odds of *someone* winning are very high. Thus, in a very large population of insects, the vast majority of the individuals may die from a severe environmental stress, but the odds are favorable for the presence of a mutation promoting the survival of a few.

With a large population, the odds of favorable mutations in the population are high. In a resource-rich environment, a reduction in the population size will trigger a positive reproduction response because the remaining individuals find themselves in an environment with very little competition among themselves for abundant resources. The growth rate of the population rises because offspring are more likely to survive in such an environment. And the greater the reduction of the population, the more pronounced the growth response in the following generation. In other words, the more intense the selective force and the more pronounced the reduction in the population size, the faster the population adapts to that selection *and* the more likely the subsequent offspring are to survive. This affects human endeavors thusly: *the more lethal* the chemical we apply to kill an unwanted insect pest, *the faster* that insect population can adapt to become resistant to the chemical.

When an intense environmental stress dramatically reduces a population and favors only those individuals with a specific mutation, the genetic variation of the population is greatly reduced. As long as that environmental stress remains strong, the mutated allele will remain at high frequency in the population. This suggests that phenotypic plasticity will not be favored in highly stressful environments because alleles for a "fixed" response will be more favorable. However, there are other factors to consider that may mitigate or reduce the loss of genetic variation.

First, the ability to produce different phenotypes under different environmental conditions (i.e., phenotypic plasticity) is in itself a genetic adaptation and not necessarily the expression of a single gene. Natural selection can therefore favor multiple genes promoting phenotypic plasticity, particularly in environments that are subject to predictable (or unpredictable) changes in growing conditions. The ability to respond adaptively to such changes will be favored over the ability to respond to a single set of conditions.

Second, in plants, if polyploidy confers a selective advantage in environments with a variety of stressors, then such species may be inherently-better able to withstand changes across a wider range of variations in the environment. Polyploid species may be predisposed to survival in those environments.

Third, if reduction in genetic variation is inevitable under conditions that cause extremely high mortality, and new mutations are the only source of new genetic variation in a population, then high mutation rates may be favored in species that experience repeated environmental extremes.

Finally, it is very likely that interspecies interactions, such as mutualistic associations among mycorrhizae, fungal endophytes, and nitrogen-fixing bacteria, provide some plants (and animals) with a buffer against the vagaries of the environment. The scope, intensity, and variation of these interactions in relation to the overall fitness of crop pests remain very poorly understood.

In short, as humans attempt to eradicate unwanted pests, it is likely that we are also selecting for suites of characteristics that favor persistence in the human-affected environment.

plasticity) because of the additional alleles and genes present, and thus we can predict that plant species, particularly weed species, displaying high levels of phenotypic plasticity could be more likely to be polyploids.

A basic assumption in evolutionary biology is that large populations will have greater genetic variation than small populations, but polyploidy in plant species implies that even small populations can express unexpectedly high genetic variation. The greater number of genetic combinations creates the potential for greater variation in phenotypic expression, but this can only be true when plants possess high *heterozygosity*; that is, the many copies of the same chromosomes have many *different* alleles for each gene. However, plant breeders and seed producers are moving inexorably toward cultivars of crop species with very low and often zero genetic variation. Combined with selection pressures on crop weeds to favor those with greater potential for plastic expression, the potential consequences for the farmer should be clear.

∼

Agriculture has created the ideal scenario for selecting wild species that can adapt to the unlimited resources of the farm. In essence, the farm scenario is the *monoculture*, an environment where only a single species of organism lives and enjoys unfettered access to essentially unlimited resources. The crop species is offered sunlight, water, and nutrients, and all competition with other species is minimized. Monocultures are essentially an artificial condition, and under only the rarest of situations are monocultures stable in natural environments. One could say that, like a vacuum, nature abhors a monoculture. The unoccupied space in a field, the unused water and nutrients, the reduction of herbivores and pathogens: all of these conditions create opportunities for other species to enter and take advantage of the ideal growing conditions. Given the thousands of plant species that have been introduced to the farms of the world, it should be no surprise that hundreds of them have found a home. And after thousands of years of farming and imposing selection pressure on the weed species, it should be no surprise that those species have become better and better at thriving in farm conditions. In essence, we have been farming weeds as long as we have been farming crops, but where humans have worked intentionally to design better crops, natural selection has worked in parallel to produce more adapted weeds for living among those improved crops.

Chapter 6

Running Faster:
Insecticide and Herbicide Resistance

The introduction of synthetic pesticides heralded a change in the practice of farming but, perhaps more importantly, also a change in our perception of how we can produce our food. Like the miracle of vaccines for the chronic sufferers of so many debilitating diseases, synthetic pesticides offered the cure for the chronic suffering of the farmer. Unlike so many cures of the past, where the cure might have been more dangerous than the disease, there seemed no downside to the use of pesticides as long as residual toxic effects could be alleviated. The promise of a more scientific approach to food and fiber production using chemistry seemed an ideal solution to the twin goals of increasing farm output and decreasing losses to crop pests.

After 60 years of steadily increasing applications and diversification of pesticides, we are faced with the fact that the goal of reducing crop losses due to pests, particularly insects, has not been achieved and does not even appear to be achievable. The pesticides used initially have been abandoned, either because they were unacceptably toxic to people or the environment or because they are were no longer effective. We face an ever-increasing list of newer pesticides entering retirement as they too lose their effectiveness, and the list of resistant crop pests grows inexorably longer. We embarked on a new and technologically advanced quest for the control of nature, but nature responded to the challenge in ways we couldn't have anticipated. After pressing the attack for many years, we now find ourselves in a position of fending off the counterattack and attempting to reduce losses.

Chemical Solutions

The evolution of resistance to the chemicals we use on the pests we are trying to control tells us something very basic: every chemical pesticide has a life span because resistance is an inevitable response to widespread use. Chemical life spans depend mainly on the rate, intensity, and distribution of use, which in turn determine the responses of the pests being targeted. However, pest resistance is also favored by other conditions, such as farming practices that encourage permanent populations of unwanted species, as is typical of perennial crops. Also, the behavior of the chemical itself is important with respect to how it is taken up by the target organism, and where and how it affects that organism. For chemical life span, a common benchmark is five years from introduction to recognizing resistance in a target organism. Unfortunately, the time needed to identify, test, and develop a new pesticide is typically eight to ten years.[1]

Every new pesticide produced, to be effective, must be chemically different from those pesticides that have become ineffective. In other words, it must have a new mechanism (or mode) of action (MOA) that affects the physiology of the target organism in a way that is lethal, yet relatively specific. The research and development branches of major chemical companies are therefore faced with the task of delivering new pesticides on a regular schedule that are capable of replacing the previous pesticides and with similar effectiveness. Hypothetically, the nearly infinite number of chemical interactions that take place inside living cells should provide an infinite number of biochemical targets for controlling the organism, but there is far more to the process.

When a pesticide is applied, it must be taken up by the target organism. In plants, for example, herbicides may be taken up through the roots as water is absorbed, or through the stomata in the leaves, or though the epidermis in different places. If through the roots, the herbicide must be applied to the soil, must be soluble in water, and must be transferred easily into the vascular system of the plant. If the herbicide is applied to the aboveground portion of the plant, it can only be taken in through the leaf stomata when they are open, usually during the day, or directly through the leaf epidermis. If the herbicide is easily degraded by sunlight, it must be applied during twilight hours. Absorption through the epidermis may be problematic because of the protective layers of wax that are typical of most plants. Regardless of how the herbicide enters the plant, it must then be translocated to the portion of the plant where it is effective.

For pesticides affecting animals, such as insects, the problems are not dissimilar. An insect must come in contact with the insecticide, which must then be taken up in some manner. Because insects are usually eating the crop plants, many insecticides target the digestive system and require nothing more than for the insect to eat or drink from the portion of the plant containing the toxin. A number of chemicals are "contact insecticides," which means that the insect must come in physical contact with the chemical. Regardless of the type of pesticide, the research and development departments of the chemical companies are faced with developing chemicals that affect different biochemical pathways in the target organism and that can be delivered easily and quickly to that site in the organism. Additional considerations concern the persistence of these new chemical products in the environment and their toxicity to non-target organisms and humans.

Herbicides

Of the herbicides currently on the market, there are only 29 MOAs in 16 different categories, even though there appear to be hundreds of different chemicals available.[2] Although each chemical company may market many pesticides by many specific trade names, those herbicides really represent only a few MOAs and no new MOAs have been introduced commercially since 1998.[3] For example, glyphosate (introduced in 1974) is a broad-spectrum herbicide that affects protein synthesis by blocking the production of aromatic amino acids such as tryptophan, tyrosine, and phenylalanine, which are required for plant growth. Any herbicide that acted in a similar fashion would fall in the same MOA, although it might have a slightly different biochemical pathway. When weeds become resistant to an all-purpose herbicide like glyphosate, they may also become resistant to other herbicides with a similar MOA. In contrast, some herbicides target particular types of plants that have specific biochemical pathways.

Both broad-spectrum and specific herbicides will eventually suffer the same fate, though in slightly different ways. Target-specific herbicides will become ineffective as resistant individuals of that plant type are selected. As the genes for resistance spread throughout the population or region, the value of that herbicide will diminish. However, this process can be relatively slow because the genetic variation for resistance must originate in that single target species or in a closely related species. Broad-spectrum herbicides (i.e., those that kill most weeds regardless of the species), tend to be used widely and frequently for all weed problems. Because many species of weeds are being attacked simultaneously, the genetic variation of a large number of species is being "sampled" and one or more genes for resistance are more likely to be present. Those genes will spread rapidly when an herbicide is used repeatedly and for nonspecific purposes. While the genes for resistance will not necessarily move between plant species, the selection of a single resistant species means that the farmer or landowner is faced with less than 100 percent weed control and an increasing problem with the resistant species. Thus, the herbicide may still be effective for most weed species, but if a resistant weed species becomes so abundant that it reduces crop production, the herbicide must be abandoned or used in combination with an herbicide specific to the resistant weed species.

The overuse, misuse, and nonspecific use of broad-spectrum herbicides has not gone unnoticed; these are important topics of discussion among farmers, researchers, pest-management experts, and the chemical industry itself.[4] In particular, the development of transgenic crops (discussed in chapter 10) allows for greater use of broad-spectrum herbicides because the crop plant is immune to the toxicity, and herbicides that kill every other kind of plant can be used without concern for any negative effects to the crop. This results in an overreliance on an herbicide with a single MOA and the likely acceleration of resistance in weed species.[5] Central to this discussion is the potential loss of the most important herbicides currently on the market, glyphosate and glufosinate, which are used worldwide and are useful for a wide range of weed problems. Both chemicals are out of patent protection and available in inexpensive generic versions, and consequently can readily be used— and *are* used—by everyone from commercial to family farmers, and from municipal to household gardeners. While both of these chemicals are becoming increasingly ineffective, the range of substitutes is very limited. In fact, glyphosate is considered a "once-in-a-century" chemical and there is no equivalent substitute on the market or nearing the market.[6]

One great concern in the agriculture industry is the loss of effectiveness in the currently available herbicide MOAs and the concurrent slowing of the development of new MOAs. As more weeds become increasingly resistant to herbicides, farmers look to the chemical industry for solutions, but few are emerging. In the past 20 years, there has been a major shift in agribusiness research toward the development of genetically modified crops and away from the development of new herbicide MOAs. In fact, the longtime leader of the industry, Monsanto, no longer synthesizes and develops new herbicides, and is now focused on biotechnology and the modification of the crops themselves.[7]

The emergence of herbicide resistance is well understood conceptually, but rather poorly understood practically. That is, we acknowledge that different species respond differently to selection pressures from herbicides, that an herbicide has different effects on different species, that different application rates influence the intensity of selection, that single mutations and multi-gene resistance are both likely, and that plants may become resistant via different mechanisms such as detoxification, sequestration, non-translocation, or non-uptake. Also, we know that resistance occurs at different organizational levels in the plant (e.g., the cell vs. the organ), and that plants can show gradations of resistance. It is also understood that the methods used by farmers for applying herbicides can accelerate or slow the emergence of resistance. In contrast, while the MOA of an herbicide can be characterized fairly specifically, the exact nature of the lethality of the MOA is not known in many cases. For example, this is the description of the biochemical action of glyphosate given by the Weed Science Society of America:

> Glycines (glyphosate) are herbicides that inhibit 5-enolpyruvylshikimate-3-phosphate (EPSP) synthase which produces EPSP from shikimate-3-phosphate and phosphoenolpyruvate in the shikimic acid pathway. EPSP inhibition leads to depletion of the aromatic amino acids tryptophan, tyrosine, and phenylalanine, all needed for protein synthesis or for biosynthetic pathways leading to growth. The failure of exogenous addition of these amino acids to completely overcome glyphosate toxicity in higher plants suggests that factors other than protein synthesis inhibition may be involved. *Although plant death apparently results from events occurring in response to EPSP synthase inhibition, the actual sequence of phytotoxic processes is unclear.*[8] [Italics added, references deleted.]

While it isn't necessary to know the exact cause of death of the plant in a biochemical sense, the evolution of resistance will be difficult to anticipate without that knowledge. Although only two mutations have been identified in glyphosate-resistant weeds, the growing dependence worldwide on glyphosate as the primary method of weed control since the introduction of glyphosate-resistant crops in 1996 has led to a rapidly increasing number of resistant weed species and biotypes.

In summary, the life span of an herbicide depends on a large number of factors, but the most important are the rates and numbers of repeated applications of a single MOA across a large agricultural region. A population of wild carrot (*Daucus carota*) was reported in 1957 to be resistant to the herbicide 2,4-D after several years of repeated application. Common groundsel (*Senecio vulgaris*) was reported to be resistant after 10 years of simazine and atrazine applications.[9] Given the current practices for herbicide use, one estimate is that resistance can occur in as few as five applications.[10] Currently, 434 biotypes of 237 weed species have been reported as resistant to 155 herbicides from 22 MOAs. Many weeds are resistant to more than one MOA, with one species (rigid ryegrass, *Lolium rigidum*) resistant to seven MOAs in Australia.[11] *Lolium* species occur in many parts of the world and, should the Australian genotype be introduced to new locations, the potential for spread or genetic contamination would be highly problematic for farmers of many crops.

Insecticides

The difficulties posed by pesticide-resistant arthropods in agriculture are not very different from those of herbicide-resistant weeds, and in some ways they are one and the same problem. Resistance in weeds and arthropods in crop fields occur for similar reasons, in the same ways, and are governed by the same principles.[12] However, where many plants can respond either genetically or by phenotypic plasticity (often due to polyploidy), animals are never polyploid and are more restricted to genetic responses to environmental stress. In addition, unlike plants, individual animals are capable of moving from one place to the next within a single generation. Thus, the effectiveness of control efforts on insect populations is influenced by the ability of the insects to disperse into and out of fields before and after insecticide applications.

There are 28 insecticide MOAs with 48 chemical types and with eight additional compounds of unknown MOAs.[13] While it is difficult to estimate the number of insecticide-resistant arthropods with accuracy, currently about 330 genera with 700 species (mostly insects and mites, not all being agricultural pests) are reported to be resistant to about 350 insecticides worldwide, and the number is rising rapidly.[14]

Perhaps to a greater degree than herbicides, the development of synthetic insecticides was limited by advancements in organic chemistry. In 1939, Paul Müller, a Swiss chemist, created the first synthetic insecticide. Dichlorodiphenyltrichloroethane (DDT), a chlorinated hydrocarbon, won Müller the Nobel Prize in Chemistry in 1948. Many chemically similar compounds were produced in the 1940s and an estimated 8 billion pounds of chlorinated hydrocarbons were used over the next 40 years.[15] These compounds were very effective general-purpose insecticidal agents and were also very persistent in the environment—which, at the time, was seen as beneficial. However, that persistence ultimately led to the widespread banning of most of the substances because of direct and indirect effects on non-target organisms and the tendency of these compounds to bio-accumulate in higher trophic levels of the food chain.

Over the next several decades, other families of insecticides were developed, proliferated, and replaced by subsequent generations of chemicals.[16] These included organophosphates, methyl carbamates, and pyrethroids. Despite high hopes and expectations for each new discovery, the majority of these compounds and their derivatives have had mixed results in terms of environmental impacts. Since the 1990s, the focus in insect control has become considerably more sophisticated. Neuroactive compounds, respiratory inhibitors, and hormone mimics that affect insect growth and development are the most common approaches. However, since the 1990s, genetic techniques that incorporate toxins directly into the crop plant have become the primary research focus. This began with the insertion of a gene for the *cry* protein from the bacterium *Bacillus thuringiensis* into the tobacco genome, which provided increased resistance to insect herbivores (see chapter 10).

Costs to Farmers and the World

Given the growing number of herbicide- and insecticide-resistant species that farmers must contend with, one could reasonably ask, Why do farmers continue to use chemicals? For example, as a result of increased weed resistance to herbicides, cotton farmers in Georgia and Alabama spend about $48 per hectare ($19 per acre) on herbicides. The answer is that once pesticides have been adopted as the dominant method for pest control, farmers have little recourse but to continue their use, and in larger quantities, with more frequent applications, and at greater cost.[17] Large-scale chemical use began in the 1950s, and the prevailing attitude was euphoric excitement over the promise of modern chemistry to eliminate diseases and pests and, therefore, human misery. But over the next few decades it became apparent that nature was more resilient than expected and that the best the chemical era could offer was short-lived and repeated alleviation of the problems.

But the treadmill was running, the race with the Red Queen had begun, and farmers had little choice but to try to keep up. The agrochemical industry has grown to mammoth proportions since 1950 and has disproportionate influence in determining the path that modern agriculture should take. Regardless of the implications and ramifications of such influence, this is an inevitable outgrowth of an unintentional race we have been running all along and have only recently acknowledged. In a very real sense, the agrochemical industry is as trapped as the farmers who buy the products.

The much larger question is whether it is possible to get off the chemical treadmill without jeopardizing agriculture, the environment, and human health. Ten years ago, the United States spent more than $10 billion annually on pest control yet sustained an additional $3 billion loss as a result of resistant pests and reduced crop yield.[18] This number is certainly greater today and will continue to grow as both the number of resistant pests and the cost of controlling them increase. The inability of synthetic pesticides to control crop pests over the long term is very clear, yet the emphasis in the chemical industry is to continue the race with the promise that a technological solution is just around the corner. In essence, the mantra of the advocates of chemical solutions seems to be "if success cannot be achieved, redefine failure."

PART III

Trying to Beat the Red Queen

If the world were so simple we could understand it, we would be too simple to understand the world.

—Emerson Pugh (paraphrased)

Chapter 7

Exercises in Futility:
Cases of Resistance

In this chapter, we will explore two particularly noteworthy examples of the problems created by resistance to synthetic pesticides, both drawn from the insect literature. The first example was selected because it *was* the first: the Clear Lake gnat, once a local problem that now epitomizes the speed of the resistance response. The second, the green peach aphid, was selected because of the scope and magnitude of the problem it presents, and because in some ways it represents the opposite end of the insecticide-resistance spectrum from the Clear Lake gnat, yet the end result is identical. Both cases are also good examples of humans learning important lessons but failing to make changes in response to those lessons. Perhaps these "exercises in futility" would be better described as *insanity*, which Albert Einstein defined as "doing the same thing over and over again and expecting different results."

The Clear Lake Gnat and DDD

The historical use of naturally occurring insecticidal substances may have resulted in resistant insects. However, it is likely that such resistance was a local phenomenon and not widespread. As of 1949, the rules of the game changed. In that year, the US Army Corps of Engineers began an experiment with a new synthetic chemical pesticide that was very specific to insects. It killed them effectively and did not appear to be toxic to organisms other than arthropods. The chemical was DDD, a chemical variant of DDT, which is a chlorinated hydrocarbon and a first-generation synthetic insecticide. DDD is somewhat less potent as an insecticide than DDT, but it is very persistent in the environment.

The test area was Clear Lake, California, a 68-square-mile (180-km²) recreational area noted for excellent weather, fishing, and sports opportunities. The target insect was the Clear Lake gnat (*Chaoborus astictopus*), a tiny non-biting fly that develops in the shallow waters of the lake in huge numbers (estimated at 640/ft²).[1] The adult gnats emerged in early summer months and made life miserable for humans attempting to enjoy the area.

In 1949, DDD was applied in a very large dose of 40,000 gallons of 30 percent DDD to Clear Lake and to 20 surrounding lakes and reservoirs within 15.5 miles (25 km).[2] The application resulted in a highly successful kill-off of the gnat, which essentially disappeared overnight and did not reappear until 1953. A second similar application of DDD was made in 1954 with apparently equal success. A final application was made in 1957, but it was largely ineffective and the use of DDD was discontinued at Clear Lake.

From this first exercise we see that two applications of DDD in five years resulted in complete resistance of the Clear Lake gnat. The enormous population of *Chaoborus* contained sufficiently high genetic variation that mutations for resistance to the chemical action of DDD were naturally present. The first application of DDD reduced the population so completely that the few remaining individuals were almost certainly naturally resistant. There were very likely preexisting and naturally occurring mutations for resistance to DDD, but in a very low percentage of the population. Such mutations do not spread initially because they do not provide a selective benefit and they may even create a physiological or energetic cost. However, once the environmental conditions changed with the addition of the stress of DDD toxicity, the carriers of the rare mutation for resistance now had a selective advantage regardless of the physiological cost associated with the mutated gene. Those individuals without the mutation suffered a much greater cost in comparison.

The population of *Chaoborus* rebounded in the four years after the first application of DDD as the surviving mutants reproduced and multiplied. Because all of the surrounding lakes were treated similarly with DDD, the likelihood of movement of nonresistant individuals from surrounding populations was greatly reduced, but such gene flow almost certainly happened to some degree. If gene flow from surrounding lakes did occur, it is very possible that several resistance mutations from the larger regional population were introduced to the Clear Lake population. The offspring of the recovering Clear Lake population were all descended from the survivors, but they may have been both heterozygous and homozygous for resistance. That is, some proportion of the individuals had two copies of a mutation, but others had only one. DDD is considered a "persistent organic pollutant" (or POP) and its effects carry over from one year to the next, but any portions of the lake with somewhat lower concentrations may have permitted heterozygotes to survive. When the abundance of the gnats fully recovered in 1953, it was probably a genetic mixture of fully and partially resistant individuals, perhaps with a very small proportion of nonresistant individuals. If the mutation for resistance carried with it an energetic cost, natural selection would still favor nonresistant individuals as the toxicity of DDD slowly declined between applications.

The second application of DDD appeared to be as effective as the first, but this was probably not the case. Large numbers of fully resistant individuals made up that population and were almost certainly unaffected. This assumption is supported by the more rapid recovery of the *Chaoborus* population in two years instead of the previous four years. The second application of DDD was as intense as the first and would have succeeded in eliminating any partially resistant and nonresistant individuals. In other words, only homozygous individuals would have survived, and in larger numbers than after the first application. The subsequent recovery was faster and comprised a greater proportion of gnats homozygous for the one or more resistance mutations. When the final application of DDD was made in 1957, very few gnats were killed; nonresistant individuals could only have been present as a result of gene flow from outside populations, and there had been fewer generations to allow for proliferation of the nonresistant genes.

The proposed solution to DDD resistance was to find a new chemical mode of action (MOA). From 1962 to 1975, a different pesticide, methyl parathion, was found to be effective at controlling *Chaoborus* and was applied annually to Clear Lake, but it also eventually lost effectiveness and was discontinued. It is very likely that the genetic variation of the original population was greatly reduced by DDD but had been renewed to some degree by mutations and gene flow by the time the methyl parathion applications began in 1962.

It is instructive to note that, despite the failure of modern chemical technology and the long-lasting ecological damage, the Clear Lake gnat is currently not the problem it once was. The annual population is lower and is thought to be held in check by the presence of two small nonnative fish (the threadfin shad, *Dorosoma petenense*, and the inland silverside, *Menidia beryllina*) that are very abundant and efficient zooplanktivores.[3] These fish represent a form of predatory bio-control of the gnat larvae that has become very effective over many years. There has been a resurgence of efforts to use biological control instead of chemical control in the past 30 years, but that focus is typically on non–crop pest species where the effectiveness of the control is measured in months and years rather than days, which is a more meaningful time frame for agricultural problems.

The ecological effects on invertebrate, fish, and bird populations from many serial chemical applications and other anthropogenic disturbances to the Clear Lake ecosystem persist to this day. The lake is considered an "impaired water body" (as defined by the Clean

Box 7-1: Hidden mutations for resistance

In animal populations, we think of each individual as representing a combination of two sets of chromosomes—one set form the paternal side and one set from the maternal side. With two copies of each gene, we can see dominant, recessive, and codominant traits, depending on the expression of the alleles of each gene. For simple genetics in which there are just two alleles for each gene (the maternal allele and the paternal allele), if we know the proportions of the two alleles in the population, we can make predictions about the makeup of the population in the next generation. For example, if the alleles are 50:50—the same number of each in this generation—we expect the same to be true in the next generation. When that is not true of the next generation, we suspect that some force, such as natural selection, has been acting on the population and changing the proportions of the alleles. With strong natural selection in favor of a particular mutated allele, that allele should become much more common very quickly.

If we start with the simplest example: two alleles for a gene (R and r) are dominant and recessive. That is, an individual possessing two R alleles has a particular trait and the individuals with two r alleles possess a different trait. For example, the RR genotype could confer pesticide resistance, while those individuals with the rr genotype would have no resistance. From basic Mendelian genetics, we know that if one parent is completely resistant (RR) and the other has no resistance (rr), all of the offspring will be Rr genotypes and could be resistant, not resistant, or, in the case of incomplete dominance, might be partially resistant. The math for that understanding also helps us make predictions about future generations and helps us understand some nuances about pest resistance.

The equation for predicting changes in proportions of these alleles in each generation is called the Hardy-Weinberg Principle and is expressed: $R^2 + 2Rr + r^2 = 1$, based on the fact that $R + r = 1$. That is, if R and r are the only two alleles, the sum of their proportions must equal 1, or 100 percent. Similarly, the larger equation equals 1 because all possible combinations of the two alleles have to add up to 100 percent. Those combinations can be RR, Rr, rR, and rr, and since Rr and rR are the same we combine them as $2Rr$.

So, if two alleles are each present in a population at 50 percent, then 0.5 + 0.5 = 1 and the larger equation is (0.5 x 0.5) + 2(0.5 x 0.5) + (0.5 x 0.5), which equals 0.25 + 0.50 + 0.25 = 1. What we predict is that RR and rr are both 25 percent of the population, or in a population of 100 individuals, 25 of them would be fully resistant and 25 would be nonresistant. If the two alleles are present in the population at 80 percent and 20 percent, then the equation yields $0.64 + 0.32 + 0.04 = 1$, which is to say we predict that RR will be 64 percent and rr will be 4 percent of the population. In a population of 100 individuals we should find approximately 64 resistant and 4 nonresistant individuals. If natural selection is favoring the R allele whether the individual is homozygous (RR) or heterozygous (Rr), then the R allele will increase in proportion in the population because those individuals are more likely to survive and reproduce. As the proportions become more extreme, the percentage of RR goes up and rr goes down very quickly. It's easy to see that the greater the proportion of one allele, the more it dominates the proportions of the next generation.

We might expect that the r allele would eventually disappear from the population as it becomes increasingly rare. However, for dominant-recessive allele situations, this may happen very slowly or not at all. When the alleles reach $R = 0.9$ and $r = 0.1$, the prediction from the equation is that 81 of 100 individuals will be RR, 18 will be Rr, and only 1 will be rr. Therefore, when the proportion of R exceeds 0.9, the prediction is that the subsequent generation will have less than 1 percent rr individuals. In small populations, in many years we might see no rr individuals at all and conclude that the r allele has been eliminated or is very nearly absent. However, just because no rr individuals are produced in a given generation does not mean the r allele is not present. First, in small populations, the r frequency could be as high as 10 percent in the population and we might not see any rr individuals out a sample of 100. Even so, we can predict that 18 of 100 individuals are Rr, which means they still carry the allele and have partial resistance to the pesticide. That is, even when we detect no nonresistant individuals in the population, the allele for nonresistance can be fairly common with nearly one out of every five individuals carrying it. And very importantly, when there are no nonresistant (rr) individuals for natural selection to eliminate, the r allele is no longer being removed from the population. When a selective stress is so intense that only homozygotes can survive it, rare alleles may be eliminated from the population, although that may still be a local rather than a regional effect.

The genetics of pesticide resistance are far more complicated than this simple example, and that makes hidden mutations even more likely to be maintained in a population. If a species is resistant to a large number of pesticides (e.g., 74 for the green peach aphid), then a large number of mutations for resistance are present and hidden in the population, as well as many more mutations that may become beneficial as environmental conditions change in the future.

Water Act), although the damage to the ecosystem extends well beyond the application of synthetic chlorinated hydrocarbons such as DDD. Nonetheless, this particular exercise in futility was the first demonstration of the concept of "bioaccumulation," and it became a fundamental message of Rachel Carson's *Silent Spring* in 1962.[4] The lesson from this first attempt to control an insect pest with a synthetic pesticide was well documented, and the clarity of the outcome should have been a stern warning about the future of chemical controls. We knew how natural selection worked, we saw it in action, and we recognized exactly what had happened at Clear Lake: we created the first pesticide-resistant superbug in an extraordinarily short amount of time. And yet the lesson learned was not that "chemical resistance happens very quickly and renders the chemical useless." Instead, the lesson apparently was that "better" chemicals needed to be developed and one can only assume the belief was that "better" pesticides would not suffer the same fate at DDD. There was no scientific basis for this belief then—nor is there now.

The Green Peach Aphid and Everything

Mutations for insecticide resistance vary in the mechanisms that protect the insect. First, for contact insecticides, mutations can reduce the permeability of the outer protective cuticle, which prevents the insecticide from being absorbed. Second, a mutation could confer greater ability to detoxify the insecticide once it has been absorbed into an insect body. Third, a mutation could alter the proteins affected by the insecticide in such a way that an insect is less sensitive to the toxicity of the chemical.[5] Thus, a variety of possible mutation pathways and combinations are possible, which can lead to resistance to insecticides in the target organisms. Any phytophagous (plant-eating) insect that can survive on a variety of host plants, under a variety of environmental conditions or under different agricultural regimes, will experience a range of selective environments that will likely favor at least one of the mutation pathways. Once that mutation is favored, the mode of reproduction can become a very important factor, too. Those organisms that are able to reproduce asexually may be at a distinct advantage, because all offspring are identical to the resistant parent.

Agricultural areas are typically either diverse mosaics of row crops or tree crops or both, or they are nearly homogeneous expanses of single crop types. Given the wide range of techniques for insecticide application, the wide range of active insecticidal compounds, and the range of dosages and intensities, there are an almost infinite number of combinations of natural selection regimes. For a widely distributed and abundant insect species, this makes the evolution of resistance to insecticides nearly inevitable.

The green peach aphid (*Myzus persicae*) has been a troublesome orchard pest for over 100 years in the United States. *Myzus persicae* is a small, soft, sap-sucking insect usually found feeding from the smaller veins on the underside of plant leaves. As large colonies build up on young leaves, the leaves curl and the plant experiences water stress and reduced growth.[6] This green peach aphid is a cosmopolitan and opportunistic feeder found on hundreds of plant species in over 50 different plant families.[7] It can be found on a wide variety of crop species ranging from tobacco to peppers, from perennial fruit trees to annual greenhouse flowers, and on many of the common weeds infesting agricultural fields.

The aphid nymphs mature very quickly, produce live young within a week of birth during the growing season, and can produce up to 80 offspring. In colder regions, the adults die in the fall and populations overwinter as eggs, but in warmer regions the aphids remain continuously active. When densities increase, winged adults are produced that will deposit eggs on nearly any plant on which they land. If these characteristics were not problematic enough, *Myzus persicae* are often parthenogenic (the female produces offspring without fertilization) in the spring and summer. For practical purposes, this means that a single dispersing individual is capable of creating a new population without the benefit of a mate.

Although physiological damage to plants and reduced yield are a problem with this pest, the much greater problem is the transmission of plant viruses. The aphid can transmit over 300 different viruses, making it the most important plant-virus vector known. The huge variety of pathogens it can transmit is related to its cosmopolitan tastes in plants and its rapid and frequent movement between so many different plant species.[8]

It should not be surprising that intense and persistent efforts have been made for decades to control the abundance of green peach aphids and the damage they cause. Because of its almost ubiquitous distribution and ability to thrive on a tremendous variety of important crop plants, this pest has been a particular target of the chemical arsenal developed in the last half century. Therefore, it should also come as no surprise that the agricultural pest that affects the most plants and transmits the most diseases also is now resistant to 75 different insecticides, by far the most of any pest species.[9] These pesticides include those from all major categories, including DDT and chlordane (chlorinated hydrocarbons/organochlorides), carbaryl and carbofuran (carbamates), diazinon and malathion (organophosphates), permethrin and cyfluthrin (pyrethroids), and imidocloprid (nicotinamides). In other words, after 60 years and multiple generations of insecticides, we are no closer to controlling this pest and quite possibly worse off, because its natural predators have been killed by the intense chemical use in agroecosystems.

The green peach aphid, one of the smallest, most innocuous, physically undefended insects imaginable, cannot be controlled with synthetic pesticides supported by the full weight of the agrochemical industry. Every attempt to curtail its explosive growth and spread using a chemical weapon has been met with a rapid counterattack in the form of genetic resistance and continued spread and devastation of our food crops. There is no better example of an organism capable of obeying the dictates of the Red Queen: run faster to stay ahead in the race. And there is probably no better example of organisms failing to recognize the reality of the Red Queen than humans using pesticides in their efforts to limit the depredations of the green peach aphid.

~

No pesticides have been officially retired due to loss of effectiveness, although many have been banned because of the large number of potential side effects.[10] Many chemicals are now of little use in the regions where they have been overused for a long period of time, but they remain in the chemical arsenal nonetheless. The potential exists that some insecticides could eventually regain some effectiveness because of the way natural selection can affect the abundance of genotypes in a population. When a farmer changes from a pesticide of one MOA to that of another MOA, the farmer is effectively changing the selective environment experienced by the crop pest. That is, resistance to the second pesticide MOA will depend on a different mutation than resistance to the first pesticide MOA. If the selective environment changes, then a mutation favored in the previous environment may no longer provide an advantage to the individual. Over many generations, with natural selection favoring only the fittest individuals, those carrying the mutation for the first MOA may be greatly reduced. Therefore, the frequency of the allele will also be reduced and the population as a whole will no longer be resistant to the first pesticide MOA. Unfortunately, given a large population of a pest species, the mutation will almost certainly be present (see box 7-1), and the return use of the first pesticide MOA will be met with a quicker resistance response than the first time it was used. Thus, although it is hypothetically possible for a pesticide to regain some effectiveness, the continued presence of the resistance mutation in the pest population will render the pesticide useless again in an even shorter amount of time than the original application.

In the case of the green peach aphid, with its widespread populations across so many host plant families and agricultural regions of the United States and the world, the size of the gene pool is enormous and incredibly diverse. Every pesticide MOA that loses its effectiveness against this enemy is almost certainly lost indefinitely as a pest-control option, especially within the time frame of the individual farmer. The loss of these MOAs will have cumulative effects on other crop species as any control efforts to contain one pest species will favor other pest species, particularly those already resistant and capable of capitalizing on the available food resource.

Chapter 8

King Cotton vs. the Red Queen

Cotton has become arguably the most important economic crop in the world in terms of tonnage and value. World production of cotton in 2011 was 29.8 million tons—twice what it was in 1980 at 15.2 million tons.[1] China is the top cotton producer, followed by India, the United States, and Pakistan, and those four countries produce more than 75 percent of the world's supply. In 2012, the United States produced 4.1 million tons or 15 percent of the total, and was by far the top cotton exporter in the world, followed by India and Australia.[2] China is usually a net importer of cotton. In some Asian countries, such as Uzbekistan and Turkmenistan, cotton production is currently the mainstay of the entire economy and can account for up to 90 percent of the economic value of all exports.

Cotton is by far the most consumptive crop in terms of the quantities of chemicals used for its production. Although estimates vary, cotton is grown on only 2.5 percent of the arable land (35M ha, 88M ac), but about $2 billion is spent worldwide on pesticides for cotton, of which the United States spends $1.3 billion.[3] This use of pesticides represents 16 percent of all insecticides in the world and

25 percent of all pesticides overall. In India, over 50 percent of all pesticides are applied to cotton, which is grown on about 5 percent of the arable land.[4] Worldwide, about 1 kilogram (2.2 lb) of chemicals is applied each year per hectare (~2.5 ac) of cotton. In 2010, upland cotton (see box 8-1) production in the United States used 45 million pounds of pesticides on 11 million planted acres, or about 4.1 pounds per acre (4.6 kg/ha).[5] These numbers have changed somewhat in recent years as more transgenic cotton has been planted (as will be discussed later).

Given the value of cotton and the growing demand for cotton products, it is easy to understand why production has escalated in the past few decades, but why has pesticide use grown even faster, considering that cotton isn't a food crop and does not produce sweet and juicy fruits for people and pests alike, yet it commands a hugely disproportionate share of the pesticides used around the world? The answer, of course, is the Red Queen.

The Simple Days of Growing Cotton

As recently as the 1940s, only three insects were considered regular problems causing serious crop loss in the southern United States, and they were well-known insects: the boll weevil (*Anthonomus grandis*), the cotton bollworm (*Helicoverpa zea*), and the tobacco budworm (*Heliothis virescens*).[6] Traditional practices such as plowing under the crop residue helped manage insect pest populations between seasons. During the growing season, insect control was managed with a few inorganic insecticides such as sulfur or calcium arsenate dust.[7] Weed control was largely performed manually by field workers wielding hoes at different times of the season. Pest control, particularly for insects, was never complete, and farmers always suffered some losses and reduced yields, but careful inspections and cheap manual labor were usually sufficient to manage the crop until harvest.

Cotton production problems were epitomized by the arrival of the boll weevil from Mexico in 1892. As the boll weevil spread and increased in abundance, in combination with bollworm and budworm infestations, cotton yields fell, sometimes dramatically, and farmers scrambled to find solutions. The entire US cotton industry was endangered, with no effective control mechanisms available. However, the development of organo-chlorine compounds as synthetic pesticides in the late 1940s appeared to solve the problem. With the spectacular effectiveness of DDT on all three major cotton pests, cotton farmers rapidly adopted chemical technology as the solution to their insect problems.[8] DDT was applied liberally in powder form, often weekly, with mechanical blowers, and was also effective at controlling many additional minor arthropod pests. Nonetheless, by the mid-1950s boll weevils were becoming resistant to DDT in many cotton-growing regions, and bollworms and tobacco budworms were quick to follow. Second-generation synthetic pesticides (organo-phosphates) supplanted the initial organochlorine compounds to control boll weevils, but they were not as effective at controlling bollworms and budworms. By the late 1950s, combinations of chemicals with different modes of action (MOAs) were being used to combat the three pests, but this approach also rapidly lost effectiveness by the early 1960s.[9] Despite the subsequent development of pyrethroids and carbamates, which suffered the same loss of effectiveness as the previous chemicals, cotton farmers sprayed and treated their fields for the boll weevil as often as every five days into the 1970s.

As chemical use became the foundation of cotton production in the 1950s and resistance became an increasing problem, farmers were forced to contend with secondary pest outbreaks. As explained in earlier chapters, a crop *ecosystem* not only has herbivores and carnivores, it is also likely to have a wide variety of them. When a crop has one or more dominant pest species, their numerical abundance may mask the presence of other subordinate and less abundant species that are largely ignored because they pose no particular threat to crop production. However, the conversion of cotton production to a chemically dependent system was centered on the use of insecticides with a specific focus on the dominant pests. While the organochlorines and organo-phosphate compounds used for insect control were broad-spectrum insecticides, they did not affect all arthropod species equally. As the abundance of boll weevils, bollworms, and budworms crashed with the application of DDT, other potential pests were released from competitive or predatory suppression or were able to invade the pest-free and very inviting fields. Thus, as the three main pests were controlled, others emerged. Over time, all species became increasingly resistant, with the result that the primary pest problems returned and were joined by new pest problems.

Unlike DDT, the newer chemicals were less persistent in the environment and had to be applied more frequently. Both older and newer insecticides affected a wide range of pests and other insects, but they were also very effective at killing many predatory arthropods and beneficial insects. When these species became less abundant, their ability to control many of the minor cotton herbivores was lost. As cotton fields were sprayed with an increasing number of toxins with increasing frequency for an increasing number of pests, the complexity and balance of the natural agro-ecosystem was greatly diminished. In a very real sense, the use of insecticides to control cotton pests created a chemical-dependency situation that seemingly could only be satisfied with the development of newer and more potent artificial insect controls because the natural controls had been eliminated.

Growing Cotton Today

The primary pests of the 1950s in the United States have mushroomed from three to more than thirty, and cotton is now the largest consumer of pesticides. Worldwide, there are an estimated 1,300 insect species and 128 species of nematodes that attack cotton.[10] The cotton industry is very aware of the issues related to production, resistance, and chemical dependency just described and has been actively involved in developing methods and solutions for managing pests. The magnitude of the problems surrounding cotton

Box 8-1: Trophic cascades

In ecological systems, diversity is manifested at every organizational level from genetic diversity to species diversity to habitat diversity. Within a community, species diversity within each trophic level is a form of *functional* diversity, which is to say that all of the ecological roles that different organisms play within the community are often well represented, even to the point of redundancy. For example, a high diversity of plant species may seem somewhat redundant because all can provide food for herbivores. However, a single species of plant can only produce abundant uniform biomass for herbivores, but 100 species of plants can produce a greater quantity of biomass and do so at different times of the year, in different qualities, at different heights, and so on. Also, some plants have deep roots and some shallow, some require more nitrogen and others more calcium, and each plant has somewhat different palatability to herbivores. This diversity at the trophic level of the producer creates niches for many herbivores, which in turn creates a diverse food base for many predators. The ability of a community to withstand and recover from severe stress, such as drought or temperature extremes, is closely related to the diversity of species within the community. Also, the regulation of the abundances of different species in different trophic levels is based on that diversity. In other words, the ecological balance is related to the diversity of plants and predators because both will regulate the abundance of herbivores, which in turn will have a reciprocal regulatory effect.

If the producer level (plants) is not restricted by resource limitations, the food base for herbivores is only limited by herbivore abundance, and their numbers will grow until the food source is strained or depleted. The abundance of herbivores will result in population growth of the predators, which will, in turn, suppress the population levels of the herbivores. When this happens, the reduced herbivore numbers will prevent the herbivores from fully exploiting their food resources. That is, the plants will recover because they are not being heavily preyed upon by the herbivores. This top-down control in systems that are not limited by bottom-up supply has been termed a *trophic cascade*.[a]

A simple example of a trophic cascade is a pond with an ample supply of nutrients. Phytoplankton (very small plants, such as algae) will colonize

a. N. G. Hairston, F. E. Smith, and L. B. Slobodkin, "Community Structure, Population Control and Competition," *American Naturalist* 94 (1960): 421–25.

the pond, and their abundance (as producers) will provide a food resource for zooplankton (very small animals). The growth of zooplankton, populations of primary consumers, will reduce the abundance of phytoplankton and the water will be reasonably clear. If small fish that prey on zooplankton are added, they represent a third trophic level—secondary consumers. As the population of small fish grows, the abundance of zooplankton will decrease, which will allow the abundance of phytoplankton to rebound, and the pond will turn green. If a piscivore such as a large predatory fish is added to a green pond with small fish, the pond water will become clear. The larger fish is a tertiary consumer at the fourth trophic level. As larger fish increase in abundance by feeding on smaller fish, the zooplankton are released from predation pressure and so their population grows. As that happens and they consume a larger proportion of the phytoplankton populations, the water clears. Thus, even though predatory fish do not eat algae directly, their presence causes the abundance of algae to decrease—this is the trophic cascade.

The protection to the balance of the community provided by consumer diversity and the trophic cascade has been lost in agricultural systems. The producer level on a large monoculture farm is extremely simple (one species), but resources are not limited and biomass production is maximized. Potential herbivores are presented with an unlimited, very uniform, and highly palatable food supply, and population growth of all herbivore species will be explosive. Under natural conditions, as herbivores thrive, predators also thrive, and this would result in eventual top-down control by the predators. In agro-ecosystems, predator species are largely absent due to toxic conditions caused by herbicides and insecticides, or because the predators are not able to overwinter in the fields and must migrate in only after they locate the prey species, which will occur only after they have reached large population sizes. Thus, the top-down control of herbivores provided by the trophic cascade is not in effect in modern agricultural fields that rely on insecticides. In essence, in such simplified communities, farmers are forced to act as the top-down control for all species of herbivores, and in modern agriculture, the top-down control mechanism is chemical.

In natural systems, each trophic level has a direct regulatory effect on those levels immediately above and below, and indirect effects on other trophic levels. The producer level (plants) ultimately influences the abundances of carnivores, and vice versa. In agricultural systems, there is (intentionally) no species diversity nor functional diversity at any trophic level, and much less regulation (if any) between trophic levels. In addition, the use of non-biological, top-down control (chemicals) is independent of and unaffected by the structure of the community or any response by other trophic levels.

production is impressive, and so, consequently, is the complexity of the solutions.

First, cotton growers were among the first to develop and adopt a modern integrated pest management (IPM) methodology, and today cotton farming and IPM are nearly synonymous in some regions.[11] The main tenet of IPM is that there is no one single answer to a pest issue; a farmer has a toolbox and should use all of the tools available rather than relying on a single method, no matter how easy or effective it appears to be. The IPM approach values insecticides as an important tool but recognizes that crop rotation, soil conservation, no-tillage practices, manual and mechanical controls, managing for biodiversity, the use of genetically modified crops, and careful timing of control practices all contribute to the control of pests, lower costs, and the health of the environment.

The implementation of IPM can be complex (and will be described further in chapter 14). For example, the University of California, Davis, has been instrumental in developing IPM guidelines, and the *IPM Pest Management Guidelines for Cotton* is over 100 pages long.[12] Very specific recommendations for control of 33 arthropod pests and six diseases cover 50 pages and are followed by 20 pages of weed-control recommendations. The recommendations are not typical of a "spray now, ask questions later" approach but instead reflect a deeper understanding of the evolutionary difficulties surrounding pest control. First, the IPM approach recognizes that the management of pests must be based on a rigorous and scientific approach that takes into account the growth stages of both the crop and the pests. Greater attention is given to the most sensitive times of the crop's growth and to the most effective times to attack the crop pests. When IPM is used successfully, the growing season is broken into very specific time periods based on the life history of cotton: preplanting to planting, crop emergence to seedling growth, budding to first bloom, first bloom to first open boll, first open boll to harvest, and harvest to postharvest. Within each time period, the concerns of the farmer are different. Some pests do not need to be managed all of the time, and greater attention is given to protecting the crop during vulnerable stages that have a greater impact on long-term productivity and yield. However, the science of IPM combines an understanding of the life history of the crop plant and the life histories of every pest, and this understanding requires an immense amount of research.

Second, the evolution of resistance to pesticides is clearly understood and is an outcome to be avoided.[13] The process leading to the development of genetic resistance is not questioned at any level from the farmer to the federal agencies with agricultural oversight. Unfortunately, the development of agricultural intensification and the infrastructure for maintaining it is predicated on the continued use of pesticides and the technologies surrounding their use. The chemical treadmill that is an integral part of most farming operations is very difficult to escape. However, as the National Research Council pointed out: "the problem is not simply that some pests *develop* resistance; some were never controlled by pesticides. For some soil-borne pathogens, nematodes, arthropods, and aquatic weeds, there are no acceptable conventional chemical pesticides."[14] Therefore, chemical solutions can only exist as one of many tools for controlling pest problems in agriculture.

Third, complete eradication of any pest is not a realistic objective and has never been achieved,[15] even though the cotton industry still has a very active program pursuing that goal with the boll weevil. Similarly, at no time in history have crop losses (for any crop) been reduced to zero, not even after the first introduction of DDT. Historical crop losses were perhaps as low as 10 percent depending on the crop (although little is known about reduction in yield); today, crop loss for cotton worldwide is estimated at 15 percent despite the advantages of modern chemistry and technology.[16] Indeed, in many places, chemicals today are used more to fend off attacks and limit losses than to make any serious attempt to eliminate the pests responsible. In India, losses from five species of caterpillars and whiteflies could potentially reduce cotton yield 20–80 percent without the protection provided by chemicals.[17] In 2012, cotton producers in the United States spent an estimated $288 million on insecticides and eradication efforts, yet cotton loss due to all factors was 39.1 percent.[18] The IPM strategy for cotton, therefore, is a more focused attempt to limit losses at critical stages of plant growth. We can only conclude that, despite all of the science and technology currently available, modern agriculture can, at best, only control crop losses to pests with no expectation of eradication.

Lastly, the overall problem is recognized as a culmination of a number of contributing factors and interactive processes: the growing number of applications of pesticides of different MOAs, intensities, frequencies, persistence, and combinations; increased mechanization; increased farm sizes; changes in our attitudes toward food and fiber production; and crop breeders and plant scientists changing the crops themselves to suit and even shape the goals of modern agriculture. For cotton, the new age of synthetic pesticides created the opportunity to focus on the development of high-yield cotton that is completely unlike the cotton of the 1940s.[19] With the use of insecticides, crop breeders could produce cotton cultivars that were much less naturally protected, more vulnerable to insect attack, and vulnerable for longer periods of time in the field. The cultivars produced more cotton fiber, but they required more fertilizer and more frequent applications of pesticides to protect the eventual harvest. Once such supercrops were developed, farmers had little recourse but to make use of these high-yield varieties that committed the farmers to a dependence on artificial fertilizers and pesticides. This technology and the accompanying costs were also exported to the rest of the world.

The Red Queen Trumps King Cotton

The story of cotton exemplifies the inevitable triumph of the Red Queen. No matter how advanced the technology (as will be discussed further in chapter 10), the ability of the pest to respond to environmental stress is essentially independent of the character of the stress. An evolutionary response is based on the process of natu-

Box 8-2: Cotton varieties

There are two primary types of cotton grown in the world, although several different names apply to each. Upland cotton (*Gossypium hirsutum*) is the dominant agricultural species and is also called American cotton or Mexican cotton. Pima cotton (*Gossypium barbadense*) is extra long staple cotton with a much longer fiber length and is also known as Creole cotton, Egyptian cotton, ELS cotton, Indian cotton, or Sea Island cotton. There are about 50 species of wild *Gossypium* around the world, and Upland and Pima cotton are cultivated hybrids of wild species.[a] Upland cotton represents about 90 percent of world production and Pima about 5 percent. In the United States, about 97 percent of cotton is Upland and 3 percent Pima cotton, of which 95 percent is grown in California and 5 percent in Arizona, where the climate is more appropriate. Of the cotton grown in California, about 40 percent is Upland and 60 percent is Pima.

In the United States in 2013, Upland cotton was harvested from 7.3 million acres with an average yield of 802 pounds per acre. Total production for the year was 12.3 million 480-pound bales.[b] Of Pima cotton, 200,000 acres were harvested, with an average 1,527 pounds per acre for a total of 634,000 bales. Of the cotton grown in the United States, only about 10,000 acres were grown as organic, pesticide-free cotton, although this sector of cotton production has grown rapidly in recent years.

---

a. Office of the Gene Technology Regulator, "The Biology of *Gossypium hirsutum* L. and *Gossypium barbadense* L. (Cotton)" (Australian Government, Department of Health and Ageing, 2008), www.ogtr.gov.au/internet/ogtr/publishing.nsf/content/cotton-3/$FILE/biologycotton08.pdf.

b. United States Department of Agriculture, National Agricultural Statistics Service, www.nass.usda.gov/Statistics_by_Subject/index.php?sector=CROPS, 2014.

ral selection, and that depends mainly on either the genetic variation within the pest population or the size of the pest population. If the population is sufficiently large, the genetic variation needed for adaptation to stress is almost certainly present. There are ways to mitigate and slow the response (see chapter 12), but this response to environmental stress is as fundamental to every living organism as any response could be.

Think of it this way: throughout the history of life on this planet, every living species on Earth has responded successfully to every stress it has ever encountered. That ability to adapt is an integral part of every species. For us to believe that one or more chemicals applied more or less haphazardly to an area, no matter how simplified, will irrevocably eliminate an insect species that has otherwise been supplied with everything it needs to survive is to ignore everything we know about evolutionary biology and ecology. From the point of view of the insect, the response to the stress is mechanical; it requires no thinking, no planning. Those individuals with the appropriate genetic code will survive, and they will perpetuate the species. This adaptation alleviates the stress, and the population recovers until the next stress exacts a response. This is the Red Queen in action.

Chapter 9

The Cornucopia of Maize
vs. the Red Queen

The reason we use pesticides is simple: to protect food and fiber production. And as the global population has mushroomed over the past 60 years, the demand for agricultural products has grown, farming intensity has increased, and pesticide use has expanded in parallel. With maize (*Zea mays*), the situation is even more complex because, particularly in the United States, protecting corn production is now less about protecting a major source of food and more about protecting the national economy.[1] While cotton is the most important fiber crop in the United States and the world, corn is the most dominant crop in the United States in terms of tonnage harvested and one that has been insinuated into nearly every aspect of the economy.

In the 1950s, corn was grown for two purposes: food for humans and food for livestock. Since then, in the United States, acreage planted to corn has increased 20 percent, from 80 million acres to almost 100 million acres, while the yield has nearly quadrupled, from 40 bushels per acre to 150 bushels/acre (1 bushel = 56 pounds). The price of corn was about $1.00 per bushel in the 1950s and hovered near $2.00 per bushel for three decades until 2005, when it began a rapid rise to over $6.00 per bushel today.[2] As a consequence, the value of the corn harvest in the United States has spiked from $22 billion in 2005 to over $80 billion in 2013. Why the dramatic increase? Because US corn is no longer grown primarily as a food source. Currently, 40 percent of the corn produced is directed to the livestock industry, but 44 percent is converted to ethanol and used in the fuel industry or exported as fuel (up from 6.27 percent in 1995). The remainder is used in myriad ways in the food and other industries.[3] Humans eat comparatively little corn directly, and corn derivatives have expanded into many different industries unrelated to food production. Through the 1970s, the main products derived from corn were sugars, starches, oils, and animal feed. The great diversification of products derived from corn now includes such seemingly unrelated products as sweeteners, organic acids, amino acids, biodegradable packaging, plastics, fabrics, coolants, cosmetics, medicines, and, of course, insecticides. The increase in demand for corn has naturally led to more intense use of pesticides and greater resistance in corn pests as a consequence.

Wild corn, or maize, is probably native to North America and is thought to have originated as a small, grasslike plant in Mexico called *teosinte* (*Zea mexicana*).[4] Native Americans selected better strains for a food crop, and it became a staple food throughout the continent. Europeans colonizing the continent adopted maize very quickly and expanded its production wherever they established settlements. By the 1800s, corn production was so important for food and feed that sections of the Midwest were known as the Corn Belt.

In the 1950s, losses to corn yields were due to three primary corn pests: the corn earworm (*Helicoverpa zea*), the European corn borer (*Ostrinia nubilalis*), and corn rootworms (*Diabrotica* spp.). Those species were responsible for about 10 percent of the losses to corn production, which were estimated at 12 percent when all other factors were included.[5] Several other insects contributed to minor losses. In 1978, the number of arthropod corn pests had risen to more than 40 for North America.[6] The numbers have grown so rapidly that corn-producing regions maintain lists of problematic species that are specific to the region. For example, the current University of California IPM Pest Management Guidelines list 30 problematic arthropods just for California corn production, as well as at least 14 (mostly fungal) diseases.[7] The UC IPM Guidelines also list 23 insecticides/miticides and 17 herbicides for use in corn, and the schedule for pesticide application is as detailed as that for cotton. Indeed, the arthropods afflicting corn in the United States have become so numerous that most lists have begun to categorize many of them by genus rather than by species.[8]

As with cotton and all other crops, as pesticide use gained a prominent role in corn production, the inevitable resistance problems arose quickly. Primary pests soon became resistant and secondary pests became problematic, additional pesticides were added to the arsenal, and increased frequency and intensity of the pesticide applications became the norm. As with cotton, a great deal of time, money, and energy had to be expended to maintain yields, to develop new chemical defenses, and to create new strategies to reduce losses in the race against the rapidly adapting crop pests.

One of the oldest strategies for managing both pests and soil quality problems is crop rotation.[9] In the United States, one of the most common crops used in rotations is soybean (*Glycine max*) because, as a legume, it can help increase nitrogen levels in the soil, and nitrogen is a macro-nutrient for plants. By rotating crops in annual cropping systems such as corn and cotton, the soybeans can reduce the need for fertilizers. More importantly, by skipping a year of corn or cotton production, the pests that specialize in corn and cotton are greatly reduced in abundance. This happens, of course, because the eggs that hatch in the alternate year produce larva with no food source available. Crop rotation essentially slows the growth of the pest population in a particular field but does not eliminate it because neighboring fields will still harbor healthy populations if they did not have a rotation year or are out of sync in the rotation. The practice of rotation is particularly important for controlling pests that live in the soil or overwinter in the soil and that are not especially mobile, such as root nematodes.

Unfortunately, any pest-control strategy used on a regular and predictable schedule is susceptible to loss of efficacy as the pests adapt. It is an evolutionary rule that predictable stresses are easier to adapt to than unpredictable stresses. In agricultural systems, the one-crop system clearly invites adaptation with its year-to-year availability of a food source for pests that favor that crop. A two-crop rotation will slow the process by alternating the availability of resources, but nonetheless pests may eventually adapt. A three-crop rotation will slow pest adaptation even more, and a rotation system that is randomized with regard to the crop could also greatly reduce some of the problems.

The predictability of the corn–soybean rotation in the Corn Belt of the Midwest in the United States was susceptible, and reports began emerging in 1987 of western corn rootworm (*Diabrotica virgifera virgifera*) damage in the first year following soybeans. Within a few years the problem had spread in the region, and in 1997 research demonstrated that some rootworm adults preferred soybeans as a location for laying eggs.[10] By adapting to a different food source for survival during the rotation year, these beetles beat the crop rotation system and became a pest on a second crop species. This "rotation resistance" was ascribed to behavioral shifts in the adult females (i.e., laying eggs on different host plants), but later research suggests a much more complex situation. A 2012 study reported that the corn rootworm larvae were feeding preferentially on leaves, not roots, of soybeans, and they produced enhanced levels of enzymes for aiding digestion.[11] This is notable because the leaves of soybeans are chemically protected from insect herbivory naturally. A 2013 study found that the natural microbial community in the gut of the rootworms showed significant shifts in enzyme production that facilitated the detoxification and digestion of the soybean leaves.[12]

Several messages emerge from this example of rotation resistance. First, the process of adaptation can involve multiple mechanisms of response, including behavioral and physiological changes, and these are genetically governed. This is not surprising, but the genetic complexities regarding simultaneous changes in multiple traits are poorly understood. The implications of such multi-trait interactions are attracting increasing attention from geneticists, evolutionary biologists, and ecologists. Second, the success of the adapting organism has reverberating effects on the other organisms affected by them, whether positively or negatively. Release from negative effects was clearly demonstrated in cotton when nondominant insects became more abundant after dominant pest species were reduced in abundance by the use of DDT and other early pesticides. However, organisms that share mutualistic relationships, as with the gut microorganisms in the rootworm, are dependent on their host for survival. Any adjustments or adaptations by the microbial community that enhance survival of the rootworm will enhance survival of the microbes. In this sense, the corn rootworm demonstrates that the Red Queen can be in action at different levels of the ecosystem simultaneously for the same species.

Third, the selection pressure imposed on western corn rootworm by the corn-soybean rotation system effectively forced the evolution of a new pest for soybeans. Under natural circumstances, it is very difficult to imagine a selective process that would have caused corn rootworms to change their food preference from corn to soybeans, which would require adaptations to overcome the protective anti-herbivory chemistry of the soy plant. However, given sufficiently strong directional selection for using soybeans as a food source, the beetles are able to survive long enough to produce eggs that can persist until the following year, when corn is once again available. This selection pressure has forced the corn rootworm to shift from being an herbivore that specializes in corn to being a more generalist herbivore of two plant species. Also, the crop rotation system forced other behavioral changes, such as changes in dispersal habits. At the end of the season, instead of laying eggs on nearby corn plants, adult corn rootworm beetles now leave the corn field in search of nearby soybean fields for egg laying.

Box 9-1: Who is the enemy?

The examples of cotton and corn, and the growing number of pest species that farmers must contend with, begs the question: How many pest species could there be? The problems with agricultural pests do not stop with the diversity of insects in one particular region. The major crops grown in agricultural areas have worldwide distribution, both in the sense that they are grown worldwide and also in that they are distributed worldwide. Many, if not most, of the pest species troubling farmers are not native but were introduced from other parts of the world. Because human activities are not restricted geographically and because human commerce is capable of moving goods very quickly from any place on the globe to any other place on the globe, all agricultural pests in the world are potential threats to every agricultural area. And because modern large-scale agriculture simplifies the agro-ecosystem to the greatest extent possible, a newly introduced crop pest is essentially presented with an unlimited resource base with few predators and very little competition. So, what is the potential scope of the problem? Let's look at the groups of organisms most likely to contribute new pest species.

First, in the phylum Arthropoda, superclass Insecta, there are about 30 orders of insects. Of those, about six orders of insects are of the greatest importance to agriculture. Estimates for the numbers of species in each order vary considerably, but Coleoptera, the beetles, outnumber all other orders combined and make up over 25 percent of all multicellular organisms on Earth. There are so many species of beetles that the British geneticist, J. B. S. Haldane, famously quipped that the Creator showed an inordinate fondness for beetles.

Order	Common name	Species (est.)
Coleoptera	beetles	700,000
Diptera	flies, mosquitoes	240,000
Lepidoptera	butterflies, moths	175,000
Hymenoptera	bees, wasps	130,000
Hemiptera	true bugs, aphids	70,000
Orthoptera	grasshoppers, locusts, crickets	26,000

Even conservative estimates predict over 1 million species in these six orders, perhaps half of which have not yet been identified. The potential number of new agricultural pest species is staggering and dwarfs the number already plaguing farmers and land managers around the world. And yet, insects might be the easiest problem to deal with because, for the most part, they are visible and largely conduct their business aboveground. The ecosystem that is the soil is considerably more complex and so poorly understood that it could be compared to a "black box" because of all the secrets it harbors. Among the poorly known soil residents are nematodes, fungi, and bacteria.

For many crops, nematodes (of the roundworm phylum, Nematoda) cause considerable damage to roots. The numbers of nematodes in soil are so great that many species are not named but identified by their feeding action; these include those that eat or feed on bacteria, fungi, protozoans, other nematodes, and plants. Soil nematodes are typically no more than about 1 mm in length, and healthy soils can harbor 100 nematodes per cubic millimeter.[a] Root nematodes can alter the development of plant roots and their ability to function, which results in stunted growth and reduced yield of the plants.

Fungal pathogens and parasites are of great concern, and the distinctions among them can be quite fine. Fungi can be external to the plant and feed on plant structures, particularly roots, or they can invade the plant and parasitize internal resources, or they can be endophytic and live entirely inside the plant. However, the relationships between plants and fungi in relation to human interests can be very complex. A fungal endophyte in *Festuca arundinacea* (tall fescue), a favored forage grass for cattle and horses, provides protection to the plant from insect herbivory but produces physiological problems for animals feeding on it ("tall fescue toxicosis").[b] The relationships among soils, plants, and fungi that lead to agricultural problems are not well understood. Particular problematic fungi include *Verticillium, Pythium, Fusarium,* and *Rhizoctonia.*

Bacterial diversity is immense, and our understanding is particularly weak regarding soils.[c] Any number of biotic and abiotic factors contribute to bacterial diversity and abundance, and the geographical distribution of bacterial types, too, is poorly understood. Although bacteria fall into a few distinct phylogenetic groupings, within those groups the bacteria are identified not by any physical structures but by the source of carbon they consume. In essence, the more different types of carbon there are, the more different types of bacteria there are. The ability of bacterial strains to adapt to changes in the environment is essentially unlimited, and almost certainly there are strains of bacteria in farmlands today that are capable of breaking down certain insecticides.

The incredible diversity of soil organisms and the severity of the problems they cause has resulted in an entirely different set of pesticides that are focused either on killing the particular type of organism (fungicides and nematicides) or just sterilizing the soil. Unfortunately, these broad-spectrum soil fumigants tend to be so toxic that they kill many non-target organisms. Many are toxic to humans as well and have been banned nationally and internationally.[d]

------

a. E. R. Ingham, "Nematodes," chapter 6 in *Soil Biology Primer,* ed. A. J. Tugel, A. M. Lewandowski, and D. Happe-von Arb (Ankeny, IA: Soil and Water Conservation Society, 2000).

b. University of Missouri Extension, "Tall Fescue Toxicosis," 2000, http://extension.missouri .edu/p/g4669.

c. N. Ferrer and R. B. Jackson, "The Diversity and Biogeography of Soil Bacterial Communities," *Proceedings of the National Academy of Sciences* USA 103 (2006): 626–31.

d. For example, broad-spectrum soil fumigants are banned by the Rotterdam Convention and the Stockholm Convention.

~

Pests and pesticide resistance in corn present tremendous obstacles for farmers and scientists, epitomizing the issues facing modern agriculture. Corn is derived from a single species, and 95 percent of all the corn grown in the United States is of a single variety ("Dent corn"—see box 9-2); moreover, the acreage is increasing. Traditional solutions to soil and pest problems, such as crop rotation, are failing. The number of pests, resistant or not, continues to increase and the number of effective pesticides continues to decrease. While the current pest-control methods are clearly limited, instituting creative, nontraditional, or integrated solutions is a monumental challenge because of the necessity of maintaining the current high production levels of corn in the United States.

Box 9-2: Corn varieties

In contrast to cotton, which has many wild relatives, corn is essentially derived from a single wild species, *Zea mexicana* or *teosinte*, and therefore different commercial types of corn represent variations on a single theme. This has very important implications for the long-term genetic health of corn, although the many land races of corn in Mexico harbor a great deal of genetic diversity to draw from. There are five commercial types of corn:

Dent or Field corn — *Zea mays indentata* is used for livestock feed, industrial products, processed foods, and all other processes not served by Flint corn. About 95 percent of the acreage in the United States is planted in field corn.

Flint or Indian corn — *Zea mays indurata* is used for similar purposes as dent corn, but is mostly grown in Central and South America and Europe.

Sweet corn — *Zea mays saccharata* and *Z. m. rugosa* are grown for human consumption, not for feed or flour. The sugar content has been bred to be much higher than in other corn types.

Flour corn — *Zea mays amylacea* is an old type of corn and is mostly used for baking because the starchy kernel is easy to grind.

Popcorn — *Zea mays everta* is another old type of corn, but with a soft starchy kernel and a very hard exterior shell.

Although each type of corn is considered a subspecies, the distinctions are largely based on sugar content, which can be modified genetically and by breeding. The different types of corn have somewhat different susceptibilities to crop pests but, for all intents and purposes, field corn is the only type grown in most areas in the United States.

Chapter 10

The Red Queen Trumps Technology: The Failures of Biotech

The history of chemical control of agricultural pests is a record of both success and failure. It is an unfortunate truth that the successes have been largely short-lived and that the original problems have never been eliminated by using chemicals—and have possibly gotten worse. The primary goal, to reduce crop losses, certainly has not been met. And the consequences, from persistent organic pollutants and secondary pest release to environmental degradation and effects on human health, have been severe. While certain harms could have been predicted and sometimes were, in the early years of the chemical revolution, some negatives were seen as regrettable but necessary collateral damage for ensuring the continued success of the agriculture industry. Whatever the issues plaguing the chemical industry, the growth of the biotechnology sector was a direct outgrowth as the struggles with pest control continued into the 1970s and 1980s. With advancements in genetic engineering, the failed promise of pure chemistry was seen as a developmental step toward real progress by biotechnology in defeating the enemies of agriculture.

Crop breeding of new cultivars is an old and respected practice in agriculture, but the new tools of the biotechnology revolution took cultivar development to new heights and in unanticipated directions. With the ability to manipulate crop genetics, no longer did the process of cultivar selection and testing have to take years and hundreds of cross-pollination trials and seed tests. The effectiveness of the insertion of a gene for a specific protein for pest or pesticide resistance could be tested almost immediately, and new cultivars could progress toward seed production, marketing, and sales in a matter of months.

With the obvious problems of pest resistance, environmental pollution, and escalating costs clearly understood, the chemical and biotech industries embarked on ambitious plans to create crop cultivars that would be genetically "immune" to the effects of the pesticides or that required no pesticide applications at all because the plants produced their own insecticidal compounds. An important creative spark came from a naturally occurring bacterium, *Bacillus thuringiensis*, or Bt, which produces a crystal protein that is toxic to insect leaf herbivores. This bacterium had been in use for some years as a natural foliar insecticide that was applied in a powdered form to the leaves of plants. For example, in western grape-growing regions in the United States, the western grape-leaf skeletonizer (*Harrisina brillians*) is a moth whose caterpillar larvae remove the photosynthetic tissues from the lower surface of grape leaves, giving a skeletal look to the leaf.[1] Control of these and other caterpillars can be achieved with applications of Bt powder sprays because the foliar herbivores ingest the bacteria as they eat the leaves. The action of the toxic protein interferes with digestion and the caterpillars quit eating and starve. Bt is very effective on a range of crop species, but its effectiveness depends on the weather, because rain will wash the powder from the leaves, and repeated applications will raise costs. Once the gene for producing the Bt crystal protein was identified and isolated, new techniques made it possible to transfer the gene (hence it can be called a *transgene*) into a chromosome of the crop plant, and in that way the plant could produce the protein and protect itself without the need for an externally applied pesticide. The same technology has been use to produce transgenic plants that are resistant to certain herbicides (e.g., Roundup Ready or RR for glyphosate resistance and Liberty Link for glufosinate resistance) so that the fields can be sprayed to kill the weeds without hurting the crop.

In 1996, Bt and RR cotton and maize were released commercially after being approved by the US Department of Agriculture. As of 2013, 90 percent of the US corn and cotton crops are genetically modified (GM) with one, two, or three traits for pest or pesticide resistance.[2] "Triple-stacked" maize (having three GM traits) was introduced in 2005, and by 2013 70 percent of all maize planted had two or more GM traits. Although current data are not available for agricultural practices around the world, estimates are that by 2009 49 percent of the cotton and 26 percent of the maize planted worldwide was genetically modified.[3] By 2010, 22 percent of the seeds sown of all biotech crops possessed two or more stacked traits.[4] The area sown in GM cotton and maize is rising rapidly and steadily each year, as is the prevalence of other GM crops, especially soybeans.

An important objective behind creating transgenic crops that are resistant to either pests or pesticides, or both, is to reduce reliance on chemicals, which are costly to buy and costly to develop, are short-lived, and are often hazardous to the environment and, sometimes, to human health. Where approved, GM crops were quickly adopted by farmers looking for relief from the difficulties of managing pests as well as managing the pesticides, which were often as technically challenging as the pests. The creation of plants with stacked traits meant that farmers could potentially dispense with both insecticides and herbicides as standard field treatments and focus their pest-control efforts on specific problems.

The shift from "external control" of insects by using insecticides to "internal control" by using plants that express the toxins is an important change in the approach to pest control. The production of crop plants that are resistant to the external controls, such as glyphosate- or glufosinate-resistant crops, is less dramatic (although still impressive) because the chemical is still being applied to the field to kill the nonresistant weeds. However, the overriding question remains the same: *Does the shift from chemicals to biotechnological "solutions" change the rules of the game relative to the Red Queen?* The answer is an emphatic NO—and there are many reasons why. A secondary question is: *Does the use of biotechnology really reduce the use of chemicals in agriculture?* In the short term, perhaps it does, but in the medium-to-long term, again the answer is NO.

Let's consider the evolutionary landscape relative to biotech crops and some other emerging issues. The traits selected for insertion into crop genomes are for specific protection from specific problems, but they do not provide a complete solution to the problem of herbivores, insect or otherwise. The Bt gene produces a single toxin that is very effective for controlling some foliar herbivores, but control of root herbivores would require a different gene insert for control. The genes for resistance to glyphosate and glufosinate protect the crop plant against damage from those specific chemicals when they are applied to control most weeds. However, many perennial weeds can be naturally resistant, because the regenerating portions of the plant, such as underground tubers, are unaffected by the chemicals. Thus, control of the diversity of potential crop pests requires a diversity of chemical or biochemical mechanisms, and this cannot be accomplished with one gene insert.

Likewise, control of a single pest species may require the production of more than one toxin because of the potential for evolved resistance to any one toxin. Increasingly, crops are being genetically engineered with stacked traits for resistance to several aspects of pest control. However, crops with stacked traits rarely have multiple protections against a single pest, but a series of single protections against multiple pests. This is important to remember. Resistance to GM crop devices, therefore, only requires a single mutation by a targeted pest to become resistant and this makes the biotechnology solution no different from that of chemical pesticides.

When we consider the ability of natural selection to counter stresses experienced by the organisms in an environment together with the diversity of organisms found at each trophic level of an ecosystem, we should also realize that the rules of the pest-control game are quite complex. Simple mutations for pesticide resistance will be favored and, given enough genetic variation (which is constantly renewed via mutations), every new pesticide will select for resistance. Essentially, every toxin selects the individuals that make that toxin obsolete. In addition, even if a pesticide is effective in the short term for a specific pest, the action of the toxin favors the population growth of any similar organism that is not affected. That is, a pesticide that reduces the abundance of a pest will favor the rapid population growth of any other pest that is not suppressed but makes use of the same resource. The problem was clearly demonstrated in cotton in the United States and China, where suppression of the major crop pests resulted in the release of previously less competitive species. Again, essentially, every toxin selects the replacement for the pest being controlled. These are the invariable rules and outcomes of the evolutionary game.

The race to beat or at least keep up with the organisms that continually threaten our crops is currently in the hands of the biotechnology sector, but the biotech race is not inherently different from the chemical race, and therefore neither are the results. The time to the first report of resistant pests can be as short as 5–6 years, and the number of resistant pests rises quickly thereafter. The difference between the two approaches is that there is a somewhat limited range of chemical modes of action and no new herbicide MOAs have been introduced in the past 20 years. The biotech industry, in comparison, is attempting to develop a highly versatile genetic palette with which to work.

The development of diverse biotech tools was prompted in part by concern over the loss of diversity of pesticide MOAs, particularly among the herbicides,[5] and the need to develop strategies for conserving important MOA.[6] Farmers are advised to adopt combinations of pest-management practices that use mechanical and biological controls when possible, to avoid reliance on a single chemical MOA, to learn more about the specific pest to be controlled and to use specific MOAs with discretion, and to use pesticides at their recommended dosage and timing. These approaches, especially in combination, are intended to slow the evolution of resistance in pest species and to conserve the value of the existing inventory of chemical controls.[7] The advent of biotechnology adds new tools to the pest-control toolbox and hypothetically reduces the risks associated with losing some of the chemical MOAs.

The development of glyphosate-resistant crops was a significant breakthrough in crop technology, but it has had some unexpected drawbacks. On the positive side, by planting glyphosate-resistant crops, farmers could wait to control weeds until after the crop had emerged, they could reduce the need for cultivation to kill weeds, they could apply herbicides fewer times, and they could practice no-till farming. All of these practices saved money and reduced the quantity of chemicals used. However, glyphosate became the herbicide of choice because it was effective on nearly all important weeds, was among the least toxic of herbicides, and within four years of the availability of glyphosate-resistant crops the patent on glyphosate expired, making it readily available and very inexpensive. As a direct result, farmers worldwide switched to glyphosate as the most effective and least expensive chemical alternative, and they began using it almost exclusively for their weed-control problems. However, the reports of glyphosate-resistant weeds began to accumulate very soon thereafter, and the life span of the "once-in-a-century" herbicide was dramatically shortened. This fate almost certainly awaits glufosinate-resistant crops as well.

With these inevitable losses of vital chemical MOAs approaching, the next biotech solution is to develop plants that are resistant to other broad-spectrum herbicides. There are several candidate compounds, but they are not as benign as glyphosate. The most likely compound for the next generation of herbicide-resistant crops will be the nonselective herbicide 2,4-D, a chlorinated hydrocarbon that was marketed in 1946 (and originally part of the formulation for

Agent Orange) and is very effective against broadleaf weeds, but not against grasses.[8] It is very likely that near-future crops will have stacked genes for multiple herbicides including glyphosate, glufosinate, and 2,4-D. This approach is expected to slow the adaptive response of the pest species because they will have to possess multiple mutations for different MOAs. However, this expectation of slowing the adaptive response is not likely to be accurate. The rapidly increasing number of species (29 in 2014) of glyphosate-resistant weeds will only have to overcome the 2,4-D treatment and, if 2, 4-D becomes a widely used substitute for glyphosate, resistance is likely to emerge within a few years. The great likelihood of this outcome is not lost on the agrochemical industry, but the business model of the industry has been co-opted by the pursuit of the Red Queen. Ultimately, the industry has no choice but to work feverishly to maintain its position and avoid losing ground in the race.

For biotech solutions to insect pests, the simple approach of inserting genes that cause plants to express insecticidal compounds has also begun to yield predictable results. Bt-resistance appeared in the United States by 2009 after the 1996 release of Bt cotton.[9] The nature of the adaptations by insects can be attributed to a variety of mechanisms for resisting the toxic effects of the chemicals expressed by the plants. For example, a mutation could slow or prevent the uptake of the toxin, change the organism's reaction to the toxin, cause the organism to metabolize the toxin faster, change the dosage required for effectiveness, or change the feeding behavior of the organism such that it is less exposed to the toxin. The number of ways for an organism to overcome the effect of a toxin is limited only by the number of possible mutations that can reduce exposure.

One of the highly touted rewards for adopting biotech solutions to agricultural problems was going to be a dramatic reduction in the use of pesticides. This prediction appears to be true in some areas and for some crops, but is not true overall. In the United States, the total quantity of pesticides used fell for the first 6 years after GM crop introductions in 1996; however, over the next 10 years, overall herbicide use rose 263,000 tons while insecticide use on Bt crops fell

61,000 tons. For the United States, the overall change in pesticide use since the advent of GM crops has been an increase of 7 percent (202,000 tons) in the past 10 years.[10] The increase in chemical use is largely due to the emergence of growing numbers of glyphosate-resistant weeds that did not exist before the introduction of glyphosate-resistant crops, coupled with the heavy reliance by farmers on that single chemical MOA. When farmers find that these weeds can no longer be controlled with low doses of glyphosate, dosage concentrations are increased more and more . . . until control with that chemical fails completely.

In China, perhaps mirroring some of the history of the United States, secondary insect pests quickly became a problem in Bt cotton fields. Attempts to control them were linked to an overall increase in insecticide use, which nearly tripled from 6.6 applications in 1999 to 18.2 applications in 2004, while the rate for non-Bt farmers was about 20–22 applications over the same period.[11] The increase in secondary insects was linked to the initial reduced use of chemicals after Bt cotton was sown. This allowed non-target insects to become more abundant and thereby precipitated an increase in the use of other chemicals. By 2004, Bt-cotton farmers in China made less money than non-Bt-cotton farmers.

Biotech Solutions and New Problems

With the advent of biotech solutions, an entirely new set of problems arose, including human introduction of resistance genes into the pest populations, unintended genetic consequences in the crop plants, and accelerated genetic erosion of crop plants.

A GM crop plant may possess a gene for resistance to an herbicide and that gene is then introduced in fantastically large numbers into the field. This is not an overstatement. Suppose 10,000 plants are sown in a field. Every cell in every part of the plant contains the resistance gene and each of the thousands to millions of pollen grains produced by every plant contains that gene. When pollen is

Box 10-1: Getting ready for the Roundup

Not long ago, the idea of transferring a gene from one organism into another unrelated organism in order to change its behavior or appearance was pure science fiction and the stuff of horror movies. In the past two decades, this process has become something that can be done in a high school science lab, albeit usually in bacteria. Today, the many biotech companies working on transgenic crops have a wide array of techniques for manipulating the genomes of the target organisms. Crop plants, in particular, are very important targets and the expression of the transgenes can make those plants resistant to the effects of the herbicide glyphosate.

Glyphosate is a molecule that is readily absorbed into plants and then interferes with the action of an important enzyme, resulting in the buildup of a molecule called *shikimate*. How this ultimately kills the plants is not clear, but the increasing shikimate prevents or interferes with other vital cellular processes. The goal behind developing glyphosate-resistant (Roundup Ready) crops was to place a gene in the plant or modify the genome such that the crop plant is less sensitive to the herbicide, while the weeds surrounding it die.

When looking for genes that produce unusual biochemical characteristics, the first and best place to look is in bacterial genomes. Bacteria are ubiquitous, reproduce rapidly, have relatively high mutation rates, and are therefore ideal organisms for generating novel mutations. The most important glyphosate-resistant gene was isolated from *Agrobacterium*, which is commonly associated with the roots of plants and is notable for its ability to transfer genes into its host.[a] As a result of the transgene, glyphosate uptake in a genetically modified plant is about 50 times lower.[b] This insensitivity to the action of glyphosate is the result of a single amino acid substitution on the affected enzyme, suggesting that naturally occurring glyphosate tolerance in weeds will undoubtedly become more common.[c]

---

a. S. O. Duke and S. B. Powles, "Glyphosate: A Once-in-a-Century Herbicide," *Pest Management Science* 64 (2009): 319–25.

b. Technically speaking, the gene doesn't confer resistance to glyphosate, but tolerance of it. That is, the organism possessing the gene becomes insensitive to the effects of the molecule.

c. T. Funke et al., "Molecular Basis for the Herbicide Resistance of Roundup Ready Crops," *Proceedings of the National Academy of Sciences USA* 103 (2006): 13010–15.

released by the crop, a field can be literally covered in copies of the transgene.

In most cases, glyphosate resistance is likely the result of natural mutations, but should the gene escape from the crop plant and somehow transfer into a weed species, genetic variation in the weed-species population in the form of mutations would not even be needed for obtaining resistance. The "mutation for success" for crop weeds would be provided by the crop itself—and there is evidence to suggest that escape of transgenes has happened on several occasions.[12] The emergence and spread of glyphosate-resistant weed species is a rapidly growing problem throughout the world, particularly in the United States. The first glyphosate-resistant weed— *Lolium rigidum*, or rigid ryegrass—was reported in 1996 in Australia and was probably a natural mutation, but the numbers have grown to 169 individual occurrences in 29 species as of 2014.[13] It is an unlikely coincidence that the numbers of glyphosate-resistant weeds grew so rapidly immediately after the introduction of crops with glyphosate-resistant genes. How many of these instances are the result of escaped genes for resistance is not known. As mentioned before, glyphosate resistance is particularly troubling because of the all-purpose nature of glyphosate, its low environmental toxicity, and the complete lack of a replacement herbicide with similar capabilities.[14]

Although we understand very well the process of mutation leading to resistance, the potential mechanisms of transgene escape from crop plants are poorly understood. The most common assumption is that transgenes are likely to move via pollen transfer between two plants that are reproductively compatible.[15] The suspected avenues for transgene escape include movement from a GM crop to a non-GM crop of the same species, between a GM cultivar and its wild version, and between a GM crop and a closely related wild species. There are other hypothetical mechanisms that, like mutations, are essentially random and very rare but could happen in populations with very large numbers. For example, a common process of inserting a gene of interest into a target organism is the use of bacterial transfer. The gene is isolated in the lab, inserted into a bacterial chromosome (plasmid), which is then "transfected" into the target cells that can then be used to generate plants with the new gene. While such transfer is understood to have happened between differing organisms, such movement of transgenes has not yet been seen in the field.[16] However, very recent research suggests that, in symbiotic systems, not only do unrelated bacteria exchange genes, but diverse species of bacteria can provide genes to their insect host's genome.[17] Given the large number of bacteria that infect and interact with plants, and the number of crop plants with which to interact, organism-to-organism movement of transgenes in the environment should come as no surprise. Such a transfer, mediated through a mutualism, would favor both the host organism receiving the new gene and the microbe, due to its dependence on the host.

Regardless of the likelihood of gene escape, even the process of inserting genes into plants has the potential for complications that are impossible to predict. Genetics, like ecology, involves complex and highly interconnected systems. Changes to a single component can have multiplicative effects, some direct and some indirect. When major genes or gene complexes are modified, it is very likely that other genes close to them on a particular strand of DNA will also be affected. This happens because a "promoter" sequence will begin the "reading" (transcription) of a gene and will continue to read subsequent genes down the DNA line until a "stop" sequence is encountered. When the activity of a gene is enhanced (up-regulated) or decreased (down-regulated), other genes that are spatially proximate to them on the same chromosome may be similarly affected. When the activity of a gene is modified, any gene that is "downstream" in terms of transcriptional processing may also be modified. Also, if a gene is up-regulated, the protein produced by that gene will be more common in the cell. If other cellular processes are dependent on the abundance of that protein, those processes will be up-regulated as an indirect effect.

For genetically modified crops, the insertion of novel genes for pest or pesticide resistance is a relatively easy process, but such manipulation may result in unforeseen effects on the phenotype of the plant (see box 10-2). For example, a 2011 experiment attempted to increase palatability of several transgenic lines of alfalfa by reducing lignin content, but the resulting plants were stunted and produced less biomass.[18] The authors suggested the research had promise, but "accompanying effects on plant development need to be better understood."

In all cases of genetic modification in crops, the end result is a genetically less variable organism because of the need for crop uniformity. Throughout history, breeding for homogeneity has been a common practice for all crops, and this has been emphasized even more in modern agriculture. A crop that grows, flowers, and matures consistently will be easier to treat for pests, a single effort will suffice for the harvest, fruit will ripen at the same time and with the same consistency, and a standard quality of produce reduces the need for culling poor or immature fruit. This uniformity saves the farmer time, energy, and labor in many ways, and it increases marketable yield. Seed producers are able to market their cultivar strains with accurate predictions for all of the above characteristics, and these cultivars tend to dominate the market. For example, two of the GM Upland cotton cultivars, Deltapine (Monsanto) and Phytogen (Dow AgroSciences), made up 84 percent of the cotton planted in the southeastern cotton zone of the United States in 2012.[19]

Crops that are highly inbred to achieve growth and production with such uniformity are developed at the cost of genetic variability. In fact, there is often no genetic variation at all. While this is seen as a positive attribute in terms of production, it contributes to the problems associated with pathogens and insect pests. If every plant in a field is the same genetically, then every plant in the field is equally susceptible to disease. Similarly, as crop species become more and more inbred and less and less genetically different, their natural chemical defenses against insects are reduced and even eliminated.[20] Both of these scenarios perpetuate the need for defensive use of chemicals and biotechnology because of the rapidity with which infections and insect outbreaks can take hold and spread. It is perhaps ironic that, while genetic variation is the key to adaptation

Box 10-2: Advanced genetic issues

All of the genes on a single chromosome are "linked" because they are all on the same strand of DNA, and the genes on one chromosome do not necessarily operate independently of genes on other chromosomes even though they aren't linked. Important metabolic functions are typically controlled in one way or another by a range of genes on many different chromosomes. It is also very possible for a single gene to influence a large number of very diverse indirect effects.

Pleiotropy is the term for *a single gene (locus) influencing multiple genes* or causing multiple phenotypic outcomes.[a] This concept is somewhat opposed to the "one gene, one protein" concept that dominated genetics for many years. Pleiotropic effects can be the result of a single gene producing different functional proteins or a protein that is used in the body in many different ways. For example, alpha and beta globins are the primary proteins that make up the hemoglobin molecule that moves oxygen through the human bloodstream. The two globins are of known lengths of amino acid chains (alpha = 141 and beta = 146), and each twists and turns into very specific three-dimensional shapes. Two alpha and two beta globins bind together to form the hemoglobin unit. If any mutation changes the order of amino acids in a globin, the 3D shape of the globin will also change and the hemoglobin configuration will change. This can easily result in a reduced ability of the red blood cells to carry oxygen. This is exactly what happens in sickle cell disease, in which the amino acid glutamine, at the sixth location on the beta globin, is mutated to the amino acid valine, and the result is an inability of the hemoglobin molecule to form the appropriate shape, which then affects the overall shape of the red blood cell. The altered red blood cells are less able to carry oxygen. This very simple *point mutation* produces a number of cascading effects that affect a range of functions in the human body. For example, a person with sickle cell disease will have a higher likelihood of experiencing anemia, ischemia, enlargement of the spleen, tachycardia, inflammation of fingers or toes, hypertension, stroke, greater susceptibility to bacterial infections, kidney necrosis and failure, and several other complications. The range of possible effects of a single gene makes the understanding of genetics very complicated as it relates to evolutionary theory.

An **epigenetic effect** is heritable *variation in gene expression* resulting in different phenotypic states that do not result from a modification of the gene, but instead are the result of mechanisms that change the expression of the gene.[b] In many cases, the changes can be as simple as the switching

---

a. F. W. Stearns, "One Hundred Years of Pleiotropy: A Retrospective," *Genetics* 186 (2010): 767–73.

b. E. Jablonka and M. J. Lamb, "The Changing Concept of Epigenetics," *Annals of the New York Academy of Sciences* 981 (2002): 82–96.

on and off of a particular gene through some biochemical interaction, and this includes what is called "gene silencing." This can be seen very clearly in developmental biology because each somatic cell contains the entire genome of the organism, but as cells begin to specialize, greater and greater portions of the genome are silenced. Eventually, only the DNA relevant to the specific function of the cell is still active. For example, a cell in the heart is only capable of functions particular to that cardiac tissue. The activation or reactivation of the silenced parts of the genome is a key to achieving some of the objectives of stem cell research.

As discussed previously, plants are frequently polyploid, which means they often have more than two copies of each allele. If we consider a diploid (2N) organism, there are three possible states of two alleles: allele 1 is expressed or allele 2 is expressed or both alleles are expressed. However, for a tetraploid (4N) organism, there are 16 possible allelic combinations if the plant is fully heterozygous. If each combination of the four alleles results in a slightly different phenotype, then polyploid plants possess the ability to adjust to a wider range of environmental conditions than do diploid plants. Individuals in a population that are capable of adaptive epigenetic expression can appear to possess a specific genotype but can also appear to exhibit strong phenotypic plasticity. For example, changes in epigenetic expression can be triggered by changes in environmental conditions, and seedlings emerging into those conditions can alter their phenotype in response, or adult plants can produce seeds that are physiologically prepared for those particular conditions.[c] The ability to alter phenotypic expression rapidly or to produce offspring with phenotypes different from the parent will be strongly favored by natural selection, especially in environments with fluctuating conditions.

Another concept that is perhaps more relevant to plant breeding is that of **gene complexes**. If sets of genes exist that together produce highly fit individuals, those sets of genes will be selected for as a group, a genetic unit, even though they are not physically linked. For example, the traits that allow a plant to survive in desert conditions affect the roots, shoots, size and shape of leaves, flowers and reproductive characteristics, and vascular systems—literally the entire organism. It is not possible for the modification of a single trait to confer the necessary physical and physiological adaptations to desert conditions. If these sets of genes have been selected together for a very long time, they may represent coadapted gene complexes in the sense that breaking them up results in a much less well adapted organism.[d]

---

c. O. Bossdorf, C. L. Richards, and M. Pigliucci, "Epigenetics for Ecologists," *Ecology Letters* 11 (2008): 106–15.

d. A. T. Ohta, "Coadaptive Gene Complexes in Incipient Species of Hawaiian *Drosophila*," *American Naturalist* 115 (1980): 121–32.

in insect and weed pests, the crops they attack are increasingly and intentionally less genetically variable and therefore less naturally able to defend themselves from attack. Of course, the GM crop breeds are genetically designed to withstand attack from specific perils, but in a very different sense they are also increasingly susceptible to such attacks.

Failure Proves the Evolutionary Rule

It is safe to say that the creation of the biotech crop cultivars has been countered by the adaptations of the targeted pests and subsequent responses of other insects in the agro-ecosystems as attempts are implemented to control the target pests. From the basic principles of evolutionary biology, these outcomes are utterly predictable. The responses of different species to environmental stress vary with the intensity of the stress and the capacity of the species to change, but the eventual outcome is always the same.

In recent years, the reports of chemical resistance in an ever-widening range of pests have grown. Some of these cases are rather amazing. For example, resistance can be to chemicals associated with the toxins rather than the toxins themselves, as in the case of cockroaches and glucose.[21] For years, cockroach bait has used glucose to induce cockroaches to consume the poison. Researchers recently reported glucose-aversion in populations of *Blatella germanica*, which appears to be an adaptive behavioral change preventing the consumption of the poisons. Although a reduced-sugar diet results in slower growth rates for the roaches, the glucose-averse individuals will be more fit if they produce more offspring than do non-averse individuals in environments with these poisons. Similarly, the western corn rootworm adults that choose to deposit eggs in soybean fields rather than on corn plants are not resistant to chemicals used in corn, but they are adapting behaviorally to changes in farming practices and these behaviors will enhance their fitness by increasing offspring success.

There are very few examples of widespread and concerted attempts at pest control that have not resulted in resistant pests. All new techniques, regardless of their provenance, that do not account for evolutionary biology will face the same result. The more specific the attempt at control (e.g., a very specific pesticide for a specific species of pest), the more specific the mutation must be to overcome the control. However, that control will only apply to the exact target organism. In contrast, broad-spectrum chemicals initially affect many different organisms, but a greater variety of mutations could mitigate the effect of the control chemical and it would only require one species to become resistant for the chemical to lose its usefulness (e.g., Palmer amaranth and glyphosate).[22] If complete control of every organism cannot be achieved, the uncontrolled organisms will become dominant, and losses in productivity or yield will continue to mount.

Environmental stress, either specific or general, favors organisms with high genetic variation, high mutation rates, phenotypic plasticity, or specific adaptive phenotypes. Any environmental stress favors individuals with at least one trait that ameliorates the effects of the stress, thereby increasing their evolutionary fitness over those individuals that do not possess such traits. And unfortunately, although we understand the process of adaptive change, what we know pales in comparison to what we don't know. It would be no exaggeration to say that the history of pest control by humans over the past 60 years is also the history of human understanding of evolutionary biology.

PART IV

Playing the Red Queen

Biological diversity is messy. It walks, it crawls, it swims, it swoops, it buzzes.

—*Paul Hawken*

Chapter 11

Understanding the Chase
to Escape the Cycle

Even on farmland, every species exists within a complex web of interactions with other organisms. These negative and positive relationships form a system that tends to dampen dramatic changes in populations. For example, if a prey species becomes very abundant, a resident predator species will grow as a result of the increased food resource. Also, other potential predators may respond by switching from a less abundant to the more abundant prey species. In the same sense that a vacuum is quickly filled by air, whenever resources are abundant or unused, the open niche space will be quickly filled by any organism that can take advantage of the resources. Such an organism gains a ready supply of food and will increase in number as its fertility rises. Optimal conditions result in a population increase of the prey species that will trigger growth in the population of its predators. But because the predator population cannot grow until the prey species has become more abundant, the change in population size of the predator will always lag behind that of the prey species. So, during the lag phase, the prey species is not being controlled by its predators.

In a highly diverse ecosystem with many species of producers (plants), consumers (herbivores), and secondary consumers (predators), the loss of one species can be balanced by the growth in number of another species. That is, the resources that were used by the missing species become available to other species, which then become more abundant. The greater the number of species, the smaller the amplitude of the variations in population size, because of increased competition. Usually, no one species is able to monopolize the unused resources. In contrast, simpler systems, such as the phytoplankton–zooplankton–small fish–predatory fish food chain, tend to exhibit a wider range of fluctuations because very few species are present at every trophic level and one species can monopolize any newly available resources. Also, in any system, the response to increased resources is dependent on the ability of each species to grow and reproduce rapidly. Smaller organisms, such as insects, tend to have very rapid growth and to reproduce in large numbers. Populations seem to explode when conditions are right, and this typically follows a pulse in the availability of resources. The larger the resource pulse, the larger the reaction from insects favored by that resource pulse. In diverse systems, with a large number of competitors present, the magnitude of the resource pulses will be smaller and will last a shorter amount of time, and population explosions will be dampened and less common. In terms of variation in population sizes, then, complex systems tend to be more stable than simplified systems.

Agriculture creates simplified systems. Monoculture farming necessarily reduces ecosystem diversity as a way to increase the resource base for the production of the desired crop species. When the producer trophic level is composed of a single species such as a crop (presumably with few weed species), the most abundant herbivores in the field will be those that specialize on that crop or that are capable of consuming it. The number of herbivore species present, therefore, is much smaller than would normally be found in a diverse ecosystem, and each species will be more abundant than in a diverse ecosystem because of the lower diversity of competitors and predators and the large resource base. Because the initial population size of the few herbivores living in a monoculture is large, the ability to produce a large rapid change in population size is greater. Predators, when present, will respond to changes in the prey population, but because of the lag time, the damage wrought by the herbivore will be greater. This problem is exacerbated when predator population sizes are already small.

Every farmer faces the task of controlling a few very abundant pest species rather than an entire ecosystem and, because of the lag time in the response of any predators living in or around the cropland, the farmer is usually forced to act preemptively or in anticipation of a pest population irruption.[1] The simpler the agro-ecosystem, the more likely this is to happen because of the small population sizes of predators and their low diversity. The use of pesticides, particularly all-purpose insecticides, will further simplify the agro-ecosystem because of negative effects on beneficial insects and other arthropods, such as spiders, which are such important insect predators. As the diversity continues to fall, a dependence on chemical controls becomes inevitable for the farmer trying to keep the herbivore pests under control. The unintended and indirect effects on species diversity are the most unfortunate and insidious problems with pesticides.

Ultimately, the practice of modern farming is not sustainable in the sense that the damage to the soil and natural ecosystems is so great that farming becomes dependent not on the land but on the artificial inputs into the process, such as fertilizers and pesticides. In large measure, the ability of crop plants to grow and remain healthy is predicated on the health of the agro-ecosystem. In a sense, the context of the ecosystem, the checks and balances, the positive and negative interactions, the ebb and flow of population cycles, all of these within and among trophic levels, have been replaced by a rigid uniformity. That uniformity has long been seen as a positive development in agriculture because of the need for dependable harvests of consistent quality with very high production levels. Modern farming technology has created a process to maximize productivity by creating the most simplified system possible, and this artificial simplicity is in complete contrast to ecological principles governing productivity in natural ecosystems.

Ecosystem ecology as a science focuses on the parameters within natural systems that influence productivity and the conditions under which the available resources are used most completely and efficiently. *Productivity* can be defined as the conversion of solar energy by plants into stored chemical energy—that is, how rapidly and efficiently plants turn sunlight into sugar. In principle, one plant species is not as efficient at capturing sunlight as two species because a second species, although very similar, will capture some of the energy in slightly different ways. Similarly, 10 species will capture sunlight more efficiently than two species because of the greater range of adaptations, structures, and capacities for intercepting solar energy and because of differences in each species' efficiency at converting that energy to sugar via photosynthesis. Fundamentally, the ecological role of plants is to convert solar energy into stored chemical energy, and there is as much variation in that process as there are species of photosynthetically active organisms. Therefore, a monoculture is not as efficient as a polyculture for increasing the total amount of solar energy captured. That is, resource-use efficiency and total productivity tend to increase with diversity.

Similarly, belowground resources such as water and nutrients are collected by each species of plant in a unique manner. The depth, size, total root surface area, spatial extent of roots, and their interactions with bacteria and mycorrhizal fungi all determine how well a plant can gather necessary resources and in what amounts. Each species has a particular rooting pattern, and therefore a combination of plants (a polyculture) with different rooting patterns will harvest soil resources more efficiently and more completely than will a single species (a monoculture) in which an individual plant has the same rooting pattern as every other individual. Ecosystems with a diversity of annual and perennial species, with grasses and broadleaf species, with summer- and winter-active species, or with evergreen and deciduous species, will use resources more completely and produce more stored solar energy than a monoculture is capable of doing.

A diverse community of plants will support a more diverse community of decomposers, soil microbes, and invertebrates. A monoculture will drop a single type of leaf and produce a single type of root, which provides a narrow range of nutrition to the soil biota. A highly diverse community produces a diverse array of resources that are heterogeneously distributed throughout the soil and across the growing season, that differ in palatability and decomposability, that differ in nutrient composition, and so on. The diversity of resources favors a diversity of consumers and reduces the dominance of any one species. As diversity in the soil community increases, so too do the speed and efficiency of nutrient recycling, and the healthier the soil will be from year to year. Soil nutrients are depleted during each growing season, but they are replenished during the winter months as leaves, stems, and roots decompose and are recycled by an active and diverse soil community. In contrast, the very simple agroecosystem dependent on a single plant species has a much simpler soil community that may become essentially inactive once the growing season ends; the depleted soil resources cannot be easily renewed for the next growing season.

A healthy, functioning community provides what are known as "ecosystem services," which are the indirect and sometimes intangible benefits that accrue from the activity of the biological components of the ecosystem. Trees transpire a huge amount of water into the atmosphere, which affects local temperatures and weather; the roots and soils slow the movement of water into rivers after rainfall, and the trees provide shelter, nesting places, and food resources for a wide diversity of animals. If trees are removed from a part of a watershed to create grazing land for livestock, for example, the forest itself is lost, of course, but all of the other related processes are also changed and perhaps lost. Ecosystem services may include good water quality and quantity many miles downstream, even during the dry months of the year. Birds that roost and nest in the forest may hunt in other habitats (e.g., owls) and can influence rodent and insect populations in nearby agricultural fields. These are all benefits of the presence of the ecosystem and they would otherwise not be available. Ecosystem services are available to farmers who maintain natural areas around their arable land and most obviously take the form of insectivores that roam out of the natural areas and into the crops in search of food. When the natural areas are degraded or converted into more cropland, those services are lost.

If the presence of plant and animal diversity can offer this wide range of benefits, and the cost of not having local diversity continues to mount, any action a farmer takes to reestablish ecosystem functions would be a positive move. Despite the damage to soils that have accumulated over the years, soil is resilient and can be restored (as will be discussed in chapter 14). Likewise, the loss of local beneficial insects and birds creates obstacles to natural pest control, but restoration of these "ecosystem functions" is possible and has been a focal point for restoration ecologists for the past 30 years and, more recently, for integrated pest manamgement (IPM) specialists as well. Just as the principles of evolutionary biology explain the rise of the pesticide-resistance problems in agriculture, so too can ecosystem ecology help us understand the necessary steps toward reestablishing an agricultural system that nurtures and makes use of natural pest controls.

∼

Our current relationship with the Red Queen and her rules is much like the gambler who has lost his initial stake in a poker game and compounds the problem by devoting more money to the game in the desperate belief that the conditions will somehow change in his favor. Despite his best effort and skill, this is a very tough challenge and one in which he starts the battle already behind. He has to win big just to come out even. To come out ahead, he has to win repeatedly for a long period of time, and that implies a change in the playing conditions that favors his strategy.[2]

Unfortunately, like all games that involve chance, the rules do not change; they are embedded in a larger context of the randomness of the draw. In the world of agriculture, both the rules of the game and that element of randomness are key. The rules of evolutionary biology are inviolable, and we must recognize and adhere to them in order to play the game. The component of randomness inherent in evolutionary biology is also governed by rules, the rules of large numbers, and in that sense we do not hold all the cards. Just as someone with five cards in poker will always lose to someone with access to ten cards, the deck is essentially stacked against us. Our opponents hold a winning hand no matter what the conditions might be.

In the pesticide-resistance game, it is necessary to adopt a different strategy; we cannot win with our current mindset. The final three chapters of this book outline a way of approaching agriculture that incorporates ecological and evolutionary principles and basically steps off the treadmill. The selected examples are not necessarily new or even innovative; they are simply methods that work with nature instead of fighting with nature. The people involved recognize that agriculture is most successful when it is approached as a partnership with the land, not a battle against it.

Chapter 12

Slowing the Race
by Slowing the Attack

Though it may seem counterintuitive, we have seen that the more effective a pesticide is, the more likely it is to promote resistance. As a rule, the more intense the killing agent (i.e., pesticide), the greater the stress on the target organism and the greater the selective pressure favoring existing genetic variation that protects the target organism from that stress. This is no more than an extension of Darwin's five postulates outlining the process of natural selection (as was discussed in chapter 3): most individuals in every population die before they reproduce, and those that survive have some adaptive trait that reduces the environmental stress below lethal levels and allows them to live long enough to reproduce. Adaptation is simply the rapid spread throughout a population of a genetic mutation that promotes survival in the face of the most potent stress. Thus, the greater the proportion of a population that a pesticide kills, the faster the resistant mutant allele will come to dominate the subsequent population, because most of the offspring of the survivors will possess the mutation. The Red Queen governs the process and is capable of shifting into hyperdrive when selective forces, such as pesticides, are very intense. The *only* way to beat the system using pesticides as the only environmental stress is to eliminate *all* individuals so that not a single resistant gene survives the chemical assault.

Therefore, the underlying core objective behind our use of pesticides is based on a flawed assumption: *eradication is an achievable goal*. It is not. In the entire history of agriculture and the long, continuous war against the pest species that consume our foods and fibers, this assumption has been disproved time and again. No important pest species has *ever* been eradicated, and the reasons why should now be very obvious. The continued attempts to increase the pressure on those species using more and more sophisticated attempts at eradication does nothing more than increase the rate at which they become resistant.

So is there a solution? In the world of food production, we play to win, never to draw or lose, and in this context winning is defined as the eradication of all competitors who attempt to "share" our food. That's our starting point, but it is an objective that is contrary to everything we have discovered about evolutionary biology and ecology. We need to reconsider; we must abandon the concept of "eradication" and replace it with the concept of "control." This is not a matter of admitting defeat, but of recognizing a simple reality: no amount of technological innovation can move us beyond the laws of physics, chemistry, and nature. Humans cannot defy natural selection; it governs the process of adaptation in all living things, from bacteria to whales.

Natural selection is embedded in the vast and complicated array of interactions between all of the organisms within an ecosystem. For example, within an ecosystem the dominance of a single organism over all other organisms is highly unusual and rarely sustained. Resource dominance by a single species brings tremendous pressure to bear on other organisms directly and indirectly affected by that dominance because of the physiological stress created by the shortage of essential resources. The result is, without fail, adaptation by natural selection to reduce that stress. Selection will favor greater specialization in the affected species as they become, by necessity, more able to carve out a narrower and more specialized role in their highly competitive environment. Ultimately, the nondominant species reclaim portions of the resource base, and the dominance by a single species is reduced.

Humans, in essence, are a species focused on resource domination, and we create intense stress on all other species in the ecosystems we influence. Those species can respond in one of two ways: adapt or die. As we know, the extinction of a population or of a species is a consequence of being unable to adapt within the time frame of the stress. However, if adaptation is possible, it will be favored. If genetic variation for resistance exists in a population, it will be selected for and will spread through the population, given enough time. As the species in the world around us adapt to our presence, they will also adapt to increase their use of the very resources we are trying to dominate.

Ultimately, our ability to produce the agricultural products we want and need is predicated on the ecological health of our agricultural environments. In many ways, our battle against the diverse array of pest species is a battle against the health of the system itself. As we kill a pest species, we also kill related species that may be beneficial, we kill predators that could assist our efforts, we reduce the ecosystem's ability to recover due to reduced diversity, and we interfere with the organisms that affect biogeochemical processes that maintain the soils in which the plants grow. By attacking a small part of nature with the goal of eradication, we limit nature's ability to provide the resources we so desire. This concept is, at its very core, the basis of "sustainability."

~

The use of highly toxic chemicals to control unwanted crop pests is not subtle. Rather, it is akin to using dynamite to catch fish: we're very sure we had an effect, but we're not always able to measure that effectiveness. Our use of high-intensity "dynamite" chemicals to attack crop pests leads inevitably to resistance and the loss of other species that could assist our efforts at pest control. By lowering the intensity of the chemical attack on agricultural pests, we could reduce and perhaps even avoid the loss of beneficial species and maintain a higher degree of connectivity within the agricultural food web. A more diverse system with greater connectivity offers humans a greater range of ecosystem services, many of which could help us control unwanted pests. The actions of beneficial insects, such as ladybugs that eat aphids, are much more subtle—they are slower yet methodical—but their effectiveness is not in doubt. And unlike highly toxic chemicals, biological controls work within the rules of natural selection: as the prey adapts, the predator counter-adapts. The rules of the Red Queen relentlessly trump our artificial attempts at pest eradication, but biological controls such as predator species are playing by the rules of the Red Queen.

By attacking crop pests with high-intensity chemical assaults, we necessarily favor very rapid, very specific solutions by the pest species to the chemical stress. A chemical attack essentially eliminates or reduces in importance all other environmental stressors. Consequently, the pest's genetic solution is specific and able to reduce the influence of the single stressor very effectively. The goal of human agriculture (i.e., complete dominance of the resource base) leads us down this path of single, high-intensity attempts at eradication. However, an approach that uses less aggressive and more diverse methods to control a pest species will slow the evolution of resistance because each stressor will be less intense and will come from different selective "directions." That is, in a diverse and highly interconnected ecosystem, every species is simultaneously contending with the natural range of existing environmental stressors such as predators, parasites, competitors, and selective removal by humans. The strong selection for a specific genotype or phenotype is not favored, and there is no single adaptive solution as there is with a dominant environmental stress. In other words, if we employed weaker but more diverse control tactics in agriculture, there would be no possibility of a single adaptive solution in any one pest species.

From a genetic standpoint, adaptation to a single environmental stress is easy and predictable. If the necessary genetic variation is present in the population in the form of a mutation for resistance, it will be favored immediately and will spread rapidly. Given that pest populations are typically not treated until they are very large, the probability of one or more mutations for resistance to a specific stress is very high. Nonetheless, although mutations are a common occurrence in large populations, they are also random with respect to future need. That is, a mutation for resistance in anticipation to a stress that has not yet occurred is not possible, and the existence of such a mutation when the stress does occur is merely a function of probability. As with the example in box 5-1, it is impossible to pick winning lottery numbers in advance, but we know that if enough lottery tickets are sold, the winning combination will exist in the population. This is a function of large numbers: even the most un-likely of events can (and will) occur, given enough time or enough opportunities. Although the vast majority of mutations are neutral or negative, given the presence of millions of mutations, at least a few of them are likely to be positive. With one ticket, my odds of winning a lottery are very small, but with a million tickets, my odds of winning are greatly improved, and the more tickets I buy, the more likely I am to hit on the winning combination of numbers. With this understanding comes the realization that all populations *depend on* mutations (i.e., genetic mistakes) for their survival in the face of uncertain environmental conditions. Even though genetic mutations are random with respect to the stressors in the environ-ment, our common assumption that mutations are "bad" does not acknowledge their absolute necessity for the long-term survival of every species.

With that in mind, if the odds of picking a winning lottery ticket are small, the odds of picking two winning lottery tickets are *ex-ceedingly* small. It is on this premise that much recent research on the control of agricultural pests is focused. By attacking crop pests from two or more different directions, the rate of emergence of

Box 12-1: How gene flow influences pesticide resistance

There are five mechanisms for evolutionary change in populations: mutation, natural selection, genetic drift, nonrandom mating, and gene flow. Gene flow is the movement of individuals (and their alleles) from one population to another, thus introducing outside genetic variation into the population. When populations are very small, gene flow can be very important for two reasons. First, small populations tend to experience high levels of inbreeding because the selection of potential mates is very limited. Second, the addition of a single individual into a small population can represent a significant increase in the number of potential mates. Thus, the new alleles can have a disproportionate effect compared to that in a large population.

In agricultural fields, large expanses of a single crop type favor large populations of specific insect and weed species. In most countries, particular regions are notable for growing particular crops exclusively (e.g., the "Wheat Belt" and the "Corn Belt"), and the entire economy is often oriented around that single crop. As a direct result, the insect pests and weed species of the region are well recognized because they are the pests for that particular crop. The pest-control practices of the region are generally adopted by everyone and are also uniform in their application. A very important consequence of this system of agriculture is that pest populations become locally adapted to the environmental stresses of that region and become genetically distinct from populations of the same species in other regions.

The evolution of local genetically distinct populations suggests that movement of the pest species into a very large agricultural region from outside populations is probably very slow because of the spatial scale. Therefore, local populations do not become diluted with outside genetic material very quickly. Second, and more important to the evolution of pesticide resistance, any genes moving in from other populations would not be adapted to the local conditions and would not be able to persist in the local population. In essence, if local populations are isolated from other populations and subjected to intense local selection pressures, they evolve into distinct genetic entities and can resist the influx of nonadapted genes from other populations. Put another way, once pesticide resistance emerges, it may be difficult to eliminate without changing the local selective environment experienced by the resistant population.

resistance will decrease dramatically, because the pest species are much less likely to possess two or more specific mutations, one for resistance to each of the stressors. The research branches of the agrochemical industry recognize that such mutation combinations should be exceedingly rare, but will eventually occur nonetheless. However, a strategy of combining stressors is one that will buy the industry much-needed time for the development of the next generation of control technology. If the time to emergence of resistance in pests can be increased from 5–6 years to perhaps 8–10 years, the hope is that the industry could spend less time and effort trying to keep up with pest resistance and focus instead on developing more-effective proactive methods of control.

~

If the single-stress pest-control techniques will invariably fail, what are the alternatives? A number of strategies have been proposed and are in different stages of testing and implementation.

Strategy 1: Use a greater variety of chemical modes of action.

Farm advisors and scientists have long advised against the widespread use of a single chemical mode of action (MOA).[1] For example, with the introduction of Roundup Ready crops, farmers have reduced the variety of herbicides and the number of herbicide applications because they can rely entirely on the use of glyphosate to control weed populations. This saves time and effort, and, very importantly, money because glyphosate is no longer an expensive patented product. Unfortunately, if all farmers in a region are planting the same crops and using the same herbicides at the same intensity (or worse, different intensities), the likelihood of weed resistance to glyphosate is not just possible, it's guaranteed, and in a very short amount of time.[2] The use of a single MOA for weed control has a predicable consequence. Instead, farmers are advised to make use of their entire chemical toolkit, even though some of the chemicals may not be as broadly effective. This could involve initial high-intensity control with glyphosate, followed by lower-intensity spot treatments with other herbicides for specific problems. Or a single herbicide could be used this year followed by a different herbicide next year, with the chemicals rotated in much the same way as the crop species. Above all else, the recommendation is to use chemicals judiciously as pest problems emerge, rather than uniformly and commonly in anticipation of the emergence of a pest problem.[3]

The calculated and cautious use of chemicals with different MOAs creates a varied selective environment for the pest species wherein no single mutation can emerge as an adaptive solution. In contrast, the increasing reliance on singular "super" chemicals not only guarantees resistance, it guarantees the rapid loss of the chemical as a useful control mechanism. The greater difficulty for the agrochemical industry is one that is not widely appreciated by the public: although the development of new pesticides typically requires 8–10 years, no new herbicide MOAs have been developed since 1998.[4] In other words, as pests become resistant to the existing arsenal of control chemicals, very few new weapons are emerging to take their place. Instead, variations on existing chemicals are produced, but the addition of a new MOA is rare. Currently, there are 28 chemical MOAs for insect control[5] and 16 MOAs (29 total variants) for weed control.[6] For both herbicides and insecticides, some are relatively specific in their action. For example, 2,4-D is a synthetic plant hormone (HRAC Group O) and is effective against broadleaf weeds but not against grasses. Several pesticides are effective against a wide range of pests. For example, DDT is an organophosphate (IRAC Group 3) that affects sodium ion channels and was originally effective against most arthropods, especially insects. In several instances, the exact biochemical effect on the target organism is unclear, particularly in herbicides. Thus, farmers and the chemical industry should work together to understand and administer pesticides in such a way as to preserve their usefulness.

Farm advisors strongly advocate for using a diversity of pesticides, in lower doses, only as needed, and across smaller scales. Unfortunately, a survey of US farmers reported in 2009 that using herbicides with multiple MOAs was one of the least-adopted "best practices" among the farmers questioned.[7] If chemicals are to remain a useful control methodology in modern farming, greater effort must be exerted to ensure their effectiveness, and the adoption of relatively simple best management practices will be an essential starting point.

Strategy 2: Use a combination of modes of action.

The previous strategy encourages the use of effective chemicals to control pests but relies on a range of chemicals to maintain control. In contrast, a second strategy is to attack pest problems by using multiple chemical modes of action simultaneously. As with the example of choosing two winning lottery tickets, the simultaneous use of two or more chemicals is predicted to prevent, or at least slow, the evolution of resistance in the pest species. That is, the use of a single highly effective chemical generates the need for a single genetic solution by creating a single dominant environmental stress. However, the use of multiple chemicals creates a multi-stress environment wherein a single mutation cannot be the evolutionary solution. The genetic solution to each individual stress is based on a single probability, but the genetic solution to all stresses simultaneously is calculated as the product of all single probabilities and consequently has a much lower probability of happening. If a mutation for each pesticide has a probability of one in a million, the probability of two mutations occurring in the same individual is predicted to be one in a trillion. That seems remotely small and highly unlikely! However, that probability would be only about 80 times smaller than winning the Powerball lottery (a one in 259,000,000 chance) in the United States if a person played it every week for a year. There are far more weeds playing the evolutionary lottery than there are Americans playing the Powerball lottery, and there are far more mutations in wild weed populations than just one per million individuals.[8]

A more insidious problem with this strategy is the general application of multiple, broad-spectrum pesticides that kill not only the target organism(s) but many more non-target organisms as well. Multiple applications of multiple chemicals will exacerbate the existing issues surrounding the loss of both ecosystem diversity and potentially beneficial organisms whether they are insects, birds, or mammals. If the simplification of our farm ecosystems weren't already a problem, such an approach to pest control might well approximate a sterilization strategy that only increases long-term degradation of the quality and resilience of farmland.

An additional problem concerning the simultaneous use of multiple chemicals is identifying which to employ. The purpose of combining chemicals is to preserve the longevity of the existing MOA while controlling the unwanted pests. Unfortunately, the chemical industry tends to combine existing MOAs that are already becoming less effective rather than introducing all-new MOAs. This approach is necessary because of the difficulty of creating new MOAs quickly, but it also illustrates the industry's failure to truly appreciate the scope of the problem. If a commonly used control is losing effectiveness, its combination with a new control mechanism does not represent a two-pronged attack, because one of the controls has already been beaten by the system. The combination merely represents a new single control mechanism with a predictable life expectancy. An example of this approach can be seen in popular flea control medications for cats and dogs. The initial generation of flea-control chemicals has long been ineffective, and the industry adopted the marketing strategy of combining the newer compounds with existing ones (e.g., Frontline with fipronil vs. Frontline Plus with fipronil and S-methoprene). This approach was only effective because fleas had not yet developed a resistance to the newer compounds. The latest products now contain three chemical compounds (e.g., Frontline TriTak with fipronil, S-methoprene, and etofenprox), with three different MOAs, but two of them already have limited success in some regions because they had been widely applied as, essentially, single-compound products. This approach toward flea control is deceptive in that it appears to be a multi-pronged attack, but is not. The end result is the serial loss of effective flea-control chemicals over a predictably short time period and the loss of the MOAs they represent. Thus, for this strategy to be effective, chemical combinations should not incorporate those that are already losing their effectiveness.

Strategy 3: Reduce the intensity, slow the response.

Most farmers would balk at the prospect of reducing the intensity of their attempts to control unwanted plant, animal, fungal, and bacterial pests because the immediate result would likely be a loss of productivity and profit from their land. Similarly, they would be hesitant to abandon the use of current chemical controls, despite understanding the problems inherent in their use and the predicted loss of long-term profitability. However, several benefits would accrue from such a shift in control tactics. First, by reducing the intensity of the selection pressure on the crop pests, their adaptive response to the pesticides would be slowed or even eliminated. A less intense approach to pest control allows other selective forces to remain in play and become more important aspects of pest control.[9] In the current climate, there is every incentive to hit the pest as hard and as often as possible in order to maintain the highest levels of harvestable productivity. The end result is obvious for all to see: escalating numbers of resistant crop pests and escalating costs of controlling the pests whether from newer expensive patented chemicals or from expensive patented seeds for genetically modified plants that possess chemical resistance.

A "slower" multi-pronged attack on pests that does not emphasize highly selective chemical stressors can reduce the rate at which pests adapt while maintaining a reasonable level of control. Merely reducing the overall kill rate on the pest species will not reduce the population size, but using several population-reduction techniques can accomplish the same goal. These are the essential objectives of the integrated pest management (IPM) strategy (chapter 14), in which chemical and nonchemical controls are employed both serially and simultaneously to replace the reliance on single mechanisms that, while effective in the short term, are destined to lose their effectiveness quickly.

Aside from slowing the development of resistance in the crop pest, other advantages include greatly reduced expenditures on expensive patented chemical products, the frequency of their application, and the energy needed to deploy them. These costs have been escalating for decades and have become one of the primary economic concerns of farmers.

Strategy 4: Increase natural controls.

A greater reliance on nonchemical solutions to crop pests will also favor the return and stability of natural predator populations. In a short time, natural predators can be reestablished as a formal control technique. Combined with a move toward reduced chemical use, the costs for pest control can be greatly reduced. The advantages of using natural predators, perhaps combined with very limited spot treatments with chemicals, should be obvious. In fact, as will be explained in chapters 13 and 14, the advantages are far greater than one would expect. The benefits of the reestablishment of a functioning, diverse, multi-trophic agro-ecosystem extend well beyond the growth of the crop plant and the maximization of the annual harvest because of the dramatic changes in the potential for long-term health of the farmland itself. This is an added bonus to the stated objectives of slowing and reducing chemical resistance in the crop pests.

Strategy 5: Reduce spatial scale, increase connectivity.

As the scale of chemical use is reduced, the ability of pests to become resistant is greatly curtailed. For example, consider an isolated field sprayed by a single insecticide. The population of insects in the treated field is under severe stress to adapt to the stress imposed by the insecticide, but the insects on adjacent properties are not. If the treated population possesses a mutation for resistance, it will normally spread very quickly through the entire population as the susceptible individuals succumb to the toxin and the mutants survive and reproduce. However, if large numbers of individuals from adjacent populations move into the (largely unoccupied) field between chemical applications, the genetics of the mutant population will be diluted and the mutation cannot become fixed in that population. Similarly, any mutant leaving the field will encounter a completely different selective environment in adjacent fields and will not be at an advantage, and the mutation will not spread beyond the original field.

The creation of a mosaic environment with different selection pressures from one field to the next prevents the success of a single dominant mutation and the eventual loss of effectiveness of useful pesticides. Also, if farmers employ a chemical rotation system as described in Strategy 1, the selective environment changes regularly and mutations in the insect population for one pesticide will likely not survive in the presence of a second pesticide. This system can only work effectively if the spatial scale of chemical application is relatively small and single pesticides are not used uniformly across entire regions.

~

Advocating a return to older, even traditional, forms of pest control may elicit knowing smiles and even outright scoffing from those on the technological side of the issue. Fair enough—let's review the most recent developments in the world of crop protection, or at least the most recent *plans* for control of pests and crop protection. And there is a difference! But first, let me add reminders of the primary obstacles facing any new technology, given the rules of the game as dictated by the Red Queen.

First, time is of the essence. It takes years to develop, test, streamline, patent, and permit a new insecticide product for public release, and it typically costs hundreds of millions of dollars.[10] More and more, the problems faced by farmers concerning the emergence of pesticide resistance occur on a shorter timeline than the production process for new chemical controls. Thus, there is a fervent and intense desire by agrochemical and agrotechnology corporations to fast-track their new discoveries. Nonetheless, as the discussion of RNA technology (below) will explain, the deployment of novel technologies can take up to 10 years.

Second, although determining the effectiveness of any new technique is part of the process of development and deployment, long-term safety is not always predictable. The companies developing new products cannot withstand a 20-year delay until the final permits are obtained. Thus, the public is asked to accept that a 2- to 3-year testing process has identified the most important short-term risks of any new product or technology. This limitation is inherent in the development of new chemicals, but it necessarily creates certain liabilities because the technology is being applied in the food production industry.

So what are the new goals and targets? The range of new ideas percolating through the industry was highlighted in a recent issue of *Science* (August 16, 2013) in a special section on "Smarter Pest Control." Pest control has entered a new era oriented around genetic manipulation of the crop species. In nearly all cases, the intent is to create *endogenous* biochemical protections (produced within the plant) that mimic the *exogenous* actions of pesticides (applied externally to the plant). The first such attempts were the creation of Bt crops that produce the *cry* protein and crops with genes providing resistance to glyphosate. These developments were deregulated by the Food and Drug Administration (FDA) in 1996. Currently, there is a spectrum of approaches for manipulating the biochemistry of the crop plant that range from the production of toxins to the production of nucleic acids that interfere with the biochemistry of the crop pest. One very important advantage of these techniques is that they can be species-specific and target only the pest species.

One particular technique that is capturing the imagination of the science world is based on the discovery of "RNA silencing" (which won Andrew Fire and Craig Mello a Nobel Prize in 2006) as a biochemical mechanism for preventing genes from being expressed. RNA molecules are small pieces of information transcribed from the DNA in the cell that serve a number of important roles. They move throughout the cell and can interact with other molecules, particularly other proteins, to produce new proteins, catalyze reactions, up-regulate and down-regulate genes, and even act to protect the cell from foreign proteins. Small interfering RNA (siRNA) are noncoding RNA that result in degradation of messenger RNA and can "turn off" (down-regulate) genes or sections of the DNA by preventing the production of particular proteins. In this way, a targeted sequence of RNA can cause a cell to go from actively producing a protein to not producing it. Such siRNAs are natural defense components of the cell and each RNA is very specific to every species.

Because it's relatively easy to design an siRNA molecule that can inactivate a protein by preventing it from being produced, the prospect now exists for "designing" a plant, for example, to produce insect-specific RNA that, once ingested by an herbivorous insect, can inactivate a vital gene in the insect's genome. This technology can potentially be tailored to target individual pest species and, in fact, can only be used in that way because the RNA itself is specific to each pest species.

Researchers in biotech corporations and universities developing new and innovative control techniques are very excited about the prospects of control through genetic engineering. In fact, this approach seems so promising that the search for chemical solutions is falling by the wayside even as the power and potential of genetic engineering is expanding. The enthusiasm in the industry is infectious, because the range of new technological tools combined with our new understanding of genetics promises a vast array of new control mechanisms that are seemingly limited only by the imagination. The scenario is hauntingly familiar: new powerful weapons with apparently unlimited potential guided by our cleverness will solve humanity's problems once and for all.

To return to the question posed in chapter 10: do the new technological advancements change the rules of the game and avoid the pitfalls that have plagued our history of pest control to this date? That is, does genetic manipulation operate outside the rules of evolutionary biology in such a way that the targeted pest species are unable to adapt to the new stress and become resistant to it? The answer is, unfortunately, no. The extant genetic variation in widespread populations of insects will eventually create resistance to the new technology. Indeed, siRNA technology appears very effective in some insect species because of the ease of delivering the molecules, but other insect species do not take up the molecules easily or the molecules are degraded before they can affect the insect. Regardless of techniques to incorporate interfering RNA molecules into the target pest species, or indeed any other method of raising selection intensity, the ability to adapt to stress is an inherent property of all living things. Because our efforts to control a biological entity interacting with another biological entity are ultimately not governed by our intentions but by evolutionary principles, the results will always be in accordance with those principles.

Despite the differences among interested parties intent on providing pest-control services to farmers, their objective is singular: eradication. Although some may pay lip service to the concept of "control," the excited language of the agrochemical industry suggests otherwise. The evidence against a "silver bullet" for pest eradication is massive, and acceptance of that reality cannot arrive fast enough. In contrast, control is very achievable, but not by using the technologies designed for eradication. Instead—and this will not be music to the chemical or biotech industries' ears—the solution is not inherently technological, and it is complex. In the following two chapters, I will make the case that the only methodology for pest control that has a realistic chance at success is adherence to the principles of evolutionary biology by reestablishing ecosystem integrity and complexity. In other words, the only way to beat the Red Queen is by playing the evolutionary game by the rules.

Chapter 13

Ecosystem Farming:
Letting Nature Do the Work

For many decades there has been a call to understand, appreciate, and employ ecosystem properties for the greater benefit of agriculture. A number of "best practices" have been adopted on a limited scale, but there are few examples of broad efforts to take advantage of the ecosystem services that support agricultural productivity. Some farmers have moved from monocultures to polycultures to increase yields; no-till farming has become a more common practice to protect soil from erosion; the use of cover crops contributes nutrients and protects soil from evaporation and heat; and the popularity of organic farming is growing rapidly, although perhaps more in response to food-quality issues than for ecosystem protection. The concept of sustainable farming is widely supported even if it has not been adopted as standard practice. Indeed, sustainable farming will never be possible without a concerted effort to integrate ecosystem principles into the practice of conventional agriculture. Nonetheless, many aspects of ecosystem farming have been implemented in many places and for many different crops.

One of the fundamental tenets of traditional farming was that of polyculture (or intercropping).[1] Around the world, on small farms and particularly family farms, different crop species were planted together and grown simultaneously in the same field. In both North America and South America, the "Three Sisters" combination of maize, squash, and beans formed a nucleus wherein the tall maize provided climbing structure for the beans, the beans provided soil nitrogen for the other crops, and the squash shaded the ground and reduced evaporative water loss. For subsistence farmers, this combination also made nutritional sense, as the three crops provide an excellent source of protein, carbohydrates, fats, and vitamins. Moreover, the natives of the American southeast were able to combine this system with inventive irrigation techniques to support a large indigenous population.[2] Other crop species and plant types, such as leguminous trees and root vegetables, were grown in addition to the Three Sisters, depending on the region and climate.

Over time, farmers in many traditional cultures moved increasingly toward plant monocultures, but this intensified approach to producing food remained relatively sustainable, as crop residues were used as secondary food sources for domesticated animals. For example, chickens, cows, and pigs could be released into the fields to forage on the remains of the crop plants, and the animals, in turn, contributed manure to the field and thus enriched the soil. This remained true in the southeastern United States until relatively recently.[3]

In eastern Asian countries, agriculture and aquaculture are often linked activities that provide both carbohydrate calories and protein for the population while maintaining an ecosystem context. In Southeast Asia, the rice field–fishery tandem incorporates traditional rice agriculture with fish harvesting in the rice paddies.[4] After the rice harvest, the flooded fields act as habitat for many species of small fishes, and this creates a protein source for the community. In China, the ancient dike-pond aquaculture system forms a complex, interwoven system that supports the production of a wide range of plant and animal resources and a very large human population.[5] The principal components of the system are fish ponds, mulberry dikes (for feeding silkworms), and sugarcane dikes. Pond aquaculture produces nutrient-rich mud that is used to fertilize both the mulberry trees and the sugarcane. The mulberry tree produces several products: the leaves are used to feed silkworms, the waste from that process is returned to feed fish in the ponds, the bark is used for paper, the branches are used as biomass or as poles for supporting other food crops. The silkworm beds are used to grow mushrooms in the off-season. Similarly, sugarcane is used to produce sugar, but the plant can also be used as feed for pigs and fish, as biomass, or as building material. Thus, the integrated agricultural system provides a diverse array of resources for the local human population while intentionally maintaining and supporting a diverse ecological system. All of this is accomplished without the use of synthetic pesticides or artificial fertilizers.

In all of these systems, agriculture is embedded in a local and regional ecological context, and its success depends on those connections. In many cases, external inputs are very important and are almost always related to water. For example, diverted river water in China renews the aquaculture ponds but also brings in sediment, nutrients, insects, plants, and even pollutants and pathogens. The system is typically able to absorb both the positive and the negative inputs because of its natural diversity. That is, the invisible bacteria, microbes, and invertebrate communities that thrive in this system are able to consume and recycle excess nutrients and minerals. They also buffer the system against outbreaks of damaging pests.

In contrast, the Nile River agriculture system of Egypt provides an important example of how the disturbance of such balanced systems can lead to systematic problems. The Nile River Delta, perhaps the world's best-known traditional system of farming, supported a very large and dense regional population in a sustainable manner for thousands of years in the midst of a vast and inhospitable desert. The farming methods were intensive, but the annual flooding of the Nile, and consequent flooding of the farm fields, deposited nutrient-rich sediments that maintained the fertility of the soils and prevented the buildup by evaporation of yield-reducing salts. Even after significant sediment deposition on farmland along the river, an estimated 60–180 million tons of sediment still reached the Mediterranean Sea via the Nile River each year. However, that movement of materials and nutrients stopped abruptly in 1964 with the completion of the Aswan High Dam.[6] Since that time, overland flooding has been completely eliminated, along with the deposition of silt and nutrients on farmland adjacent to the river. (However, the reservoir of water behind the dam and its metered release has allowed for a dramatic increase in the acreage of land under cultivation.) It is also estimated that about 60 percent of the Nile water evaporates after leaving the dam, which increases the natural salt concentration of the river.[7] With the saltier river water, the high rate of evaporation in the Egyptian sun, and the lack of annual flood waters to remove salt buildup in the soil, by 1982 half of all irrigated cropland was threatened by salinization.[8] Of greater importance to this discussion is the heavy dependence on pesticides in Egypt today and the complete loss of sustainable agriculture along the Nile River. Egypt is now Africa's largest consumer of pesticides, most of which are used for cotton production, and the Nile River now carries a heavy pesticide and fertilizer load to the Mediterranean Sea (some of which originates in countries upriver from Egypt). While many factors affect agriculture in Egypt, the simplification of the Nile agro-ecosystem that has occurred in the 50 years since the building of the Aswan High Dam is inextricably linked to the need for greater artificial inputs from fertilizers and pesticides. The dependence on artificial inputs is self-perpetuating as the entire system becomes less connected to the natural processes that favor ecosystem diversity.

Agricultural practices can support ecosystem services and stability by maintaining a greater variety of niches for insects, birds, and other beneficial species through polycultures, high crop diversity, and greater area devoted to wildlife habitat. While polycultures and adjacent habitats may provide refuges for herbivorous insects, the presence of predatory and other beneficial species typically greatly outweighs the negative consequences of the move away from

monocultures and "clean" farming. The presence of insect predators within a more diverse local ecosystem buffers the larger system
from large-scale population irruptions of harmful species and reduces the need for chemical interventions.

A very common farming system that is intermediate between
polyculture and monoculture is the practice of crop rotation in
which a particular field is not planted with the same crop in suc-
cessive years. This approach is widely practiced in the American
Midwest, where maize and soybeans form a long-standing two-crop
rotation. (See chapter 9.) This practice was adopted for two specific reasons, both of which are unrelated to either maintaining local
ecosystem diversity or practicing sustainable farming. Soybeans, as
a legume, can contribute to soil fertility due to the nitrogen-fixing
bacterial colonies in their roots. More importantly, as the pest species attacking maize have become so problematic and chemical control so expensive (and often ineffective), a low-tech and economical
solution has been to adopt a crop-rotation system that suppressed
maize pests while still producing a commodity crop. The insects and
other pests that attacked maize were unable to persist in large numbers if an insect-resistant crop (such as soybeans) was planted in alternate growing seasons. However, for the American Midwest, the
two-crop rotation is hardly different from the monoculture of maize
in terms of the overall biodiversity of the region.

Crop rotation systems making use of a large variety of crops create a matrix of habitats that change dramatically from year to year.
By itself, this creates a more diverse regional ecosystem, but unless
one or more of the crops is a perennial species, the entire matrix
disappears at the end of each growing season as the crops are harvested. If winter crops are not sown, the months between warm-
season crops are characterized by immense acreages of bare, fallow
soil. Only those animal species able to overwinter in the soil or in
the crop residue can persist in such a system unless they are able
to migrate to permanent habitat nearby. Larger animal species that
require some degree of vertical vegetation structure to provide suitable habitat (e.g., for roosting and protection) are not able to persist

Box 13-1: Jack of all trades or master of one?

All species, whether herbivores or carnivores and even plants, can be either habitat specialists or generalists. A generalist species is capable of using a wide range of resources, while the specialist is adapted for exploiting one particular resource. A generalist species is less dependent on any specific resource for survival, while a specialist can monopolize a specific resource and exclude generalists. As a rule, invasive species tend to be generalist species that move easily and across long distances, can find needed resources relatively easily, and can survive because they do not require another particular species to be present in the environment, such as a particular food source in the case of animals or a specific pollinator in the case of plants. Many introduced crop pests are generalist species that are able to establish very quickly and reproduce in large numbers, and this is particularly true of weedy plant species.

However, agricultural animal pest species are often specialists. They are dominant species because they can thrive on a specific crop plant, often to the exclusion of other pest species. To do this, they usually display very rapid growth rates and very high fertility, are mobile within the specific resource (the crop), and consume the resource very quickly. Other species that cannot match these characteristics can be excluded because they are simply outcompeted for the resource and crowded out of the resource space. A pest species that specializes on a particular crop can consume the resource faster than the nonspecialist species. Thus, because modern farming produces monocultures of certain crop plants, specialists are favored if they can quickly dominate that food resource and outcompete the generalist species.

Decades of monoculture production coupled with pesticide use has favored a range of very particular pests for each crop species. Each of the specialist pests is capable of very effectively attacking a different part of the crop plant and at different stages of the growing season. For example, particular insect species feed on stems, leaves, buds, flowers, or seeds. Consequently, the farmer is not faced with controlling one species or applying a chemical at one time of the growing season, but increasingly is confronted with many pest species and has to apply many different chemicals at many different times.

The specialist pests do not interact directly and do not compete for the same portion of the crop plant's resources, but they are, as a group, capable of attacking all or most parts of the crop plant simultaneously or sequentially. Indirectly, the pests compete with each other because if the leaves are eaten by a leaf specialist, the plant may produce fewer flowers for the seed specialist. However, the overall result can be a complete destruction of the crop in terms of productivity.

The concept of generalist vs. specialist raises questions about the use of biological control agents—for example, a predatory insect to control a herbivorous insect. Is it better to find generalist or specialist biological controls of the pest species? Typically, the goal is to find specialist control species for specific pest species in the sense that the control species preys upon the specific pest species and only that pest species. If a control species, such as a predator, is imported from another country and does not have a very specific prey, the potential exists for the predator to become a nuisance itself by proliferating and eating other prey, such as native species. Therefore, intense scrutiny and research are necessary to ensure that the new import eats only the species it is intended to control without also preying on other potentially beneficial or rare native species. Unfortunately, this goal of prey specificity may be a daunting task because it is often necessary to search long and hard to identify a specialist predator. Additionally, there is nothing to prevent the introduced predator from adapting to the new environment and finding non-target prey species.[a]

If a predator is a specialist and can eat only a single prey species, it becomes impossible to maintain high population levels of the predator once it has begun to control the prey species. In fact, the more successful the predator is, the less food there will be, and the more the predator population will shrink. As the predator population shrinks, its control of the prey species will also shrink with the result that the prey species will begin to rebound. Of course, as the prey species population recovers, the predator will also recover, but at a slower pace. It is a corollary of predator–prey dynamics theory that a specialist predator will not eliminate its prey in an open system because of this cyclical population behavior. As the prey population grows, the predator population will grow in response, but that will drive down the prey population, which results in a decline in the predator population as its food resource disappears. The decline of the predator population allows the prey species to recover, and so on. Occasionally, due to random fluctuations, a very small predator population can disappear entirely and has to be reintroduced to avoid a catastrophic population explosion of the prey species.

In contrast, a generalist biological control species is able to switch between prey species and focus on the most abundant or eat any species it encounters. The predator will consume the prey species in proportion to their population sizes and no one species will be able to grow to large numbers. By having multiple prey species to attack, the population size of the generalist predator

---

a. See, for example: R. L. Koch, "The Multicolored Asian Lady Beetle, *Harmonia axyridis:* A Review of Its Biology, Uses in Biological Control, and Non-Target Impacts," *Journal of Insect Science* 3 (2003): 32, www.insectscience.org/3.32/.

remains higher and does not experience dramatic fluctuations as does the population of the specialist predator. In this way, a generalist predator can serve as a control for many species, and there is much less concern over its disappearance due to food shortages causing very low population sizes.

The promise of biological control of pest species with either specialist or generalist predators has generated books and journals dedicated to understanding the science and the application. However, although biological control is a tremendous improvement over attempts at chemical control, neither is capable of eradicating the targeted pest species. In addition, the reduction of the prey species population by a predator creates an opportunity for any non-prey pest species to dominate the now-available crop species. This is the nature of the secondary pest problems described for cotton in chapter 8. Thus, the importation of a predator for control of a pest is, in some ways, analogous to the use of a single chemical MOA for control of a particular pest species in that it seeks to address a single component of a very complex system with no way to account for the subsequent responses of the system.

amidst annual crop rotations, and the local ecosystem is dominated by small species, particularly insects, that are adapted to such highly and regularly disturbed conditions.

A diverse ecosystem with its multiplicity of interconnected functions, operating at a natural pace and scale, is essential to sustainable farming. This concept is the foundation of Masanobu Fukuoka's farming philosophy in his book *The One Straw Revolution*.[9] His four tenets of "natural farming" very clearly opposed the use of artificial mechanisms to promote rapid and unnatural growth or to reduce the presence of unwanted species. These tenets of natural farming were as follows:

1. Maintain healthy soil. Do not damage the soil with cultivation (plowing).
2. Promote the natural fertility of the soil. Do not attempt to enhance the productivity of the crops with chemical fertilizers.
3. Use the natural productivity of the field to control unwanted plants. Weeding by tillage or herbicides only encourages dominance of unwanted species.

ecosystem Farming: Letting Nature Do the Work 167

4. Diversity is the best way to control pest species. A dependence on chemicals for insect control will only create additional problems.

Although Fukuoka was best known for his approach to rice farming in Japan, his farms included a wide range of annual crops such as daikon (radish) and perennial crops such as oranges, and he applied these principles to all crops.

When reading Fukuoka's writings, it is apparent that his greatest frustration with those he was trying to convince was a particular stubborn obstacle characteristic of contemporary thinking: the notion that the best approach for dealing with a problem is direct action using available modern technology rather than waiting for nature to take its course. To do otherwise is analogous to having a serious infection, but eschewing a potent antibiotic in favor of the body's immune system. In contrast, Fukuoka argued that the use of artificial supports does not strengthen the ecosystem, and chronic use inevitably weakens it. However, the desire for rapid and effective responses to problems has been fueled by technological advances that appear to provide tremendous control over nature to the benefit of humankind. This kind of power is difficult to abandon.

In Fukuoka's system of farming, an imbalance created by, for example, a large leafhopper population did not demand an instantaneous response with chemicals from the farmer, but patience as the balance was regained naturally because of the responses of resident predators and competitors. Weeds were allowed to grow; diseased plants were not removed from the field; yields were always as high, or higher, than neighboring conventional fields. Fukuoka spent a great deal of time observing the soil and organisms other than those he was farming because the health of all of the components of the farm was the foundation for the production of a healthy crop. More importantly, Fukuoka insisted, and demonstrated, that such an approach was essential to the long-term health of the agro-ecosystem and was the best, if not the only, way to farm sustainably.

As we will see in chapter 14, the resilience of soil and plant and animal communities suggests that it may not be too late to

Box 13-2: Living soil

It may be an exaggeration to view plants as an epiphenomenon of soil, but perhaps not. Our most complete understanding of plant growth is aboveground, and we truly know very little of what there is to know about what happens belowground. A healthy topsoil may contain, both by mass and by volume, less "dirt" (sand, silt, and clay) than it does air, organisms, and organic compounds. Just as every cubic milliliter of seawater may contain 10^7 viruses and 10^5–10^7 bacteria of perhaps 10^3–10^5 different "species,"[a] so too does soil contain a fantastically rich array of living organisms, all interacting in a complex food web. Listed below are the different biotic components of soil, numerical estimates of each in a handful of soil, and estimates of biomass per acre.[b]

Type of organism	Abundance	Biomass (lb/ac)
Plant roots	10^3 (inches)	10,000
Archaea and Bacteria	10^8–10^{10}	100–1,000
Actinomyces	10^8–10^9	100–1,000
Fungi	10^5–10^8	100–1,000
Protozoa	10^5–10^7	10–100
Nematodes	10^3–10^4	10
Micro-invertebrates	10^2–10^3	1–10

All of these components should be viewed in the same food web and ecosystem context as aboveground biota: they all have roles as herbivores, predators, pathogens, parasites, detritivores, and decomposers. Many of their interactions are via direct contact, but a very large number are chemically mediated. Their activity levels together and within different components are also mediated by moisture, temperature, pH, oxygen availability, and time of year.

In a healthy soil, as with any ecosystem, the different organisms are balanced from year to year with predators regulating numbers of prey, resource consumption efficiently matching resource availability, and all components

---

a. University of Bergen, Department of Marine Microbiology, "Viruses and Bacteria," www.uib.no/rg/mm/artikler/2009/01/viruses-and-bacteria (last updated June 2010).

b. Michigan State University, Department of Crop and Soil Sciences, "Soil Ecology and Management: Soil Biology," www.safs.msu.edu/soilecology/soilbiology.htm (2004).

varying in abundance with the seasons. When soil is disturbed or treated by a fungicide that eliminates most fungi or a nematicide that eliminates nematodes, entire components can be eliminated or greatly reduced and their function in the soil ecosystem lost or diminished. The organisms higher in the food chain that fed on them will be strongly affected. More importantly, the organisms that were regulated by the missing component can become unregulated and expand numerically, and greatly influence the activities and abundances of the rest of the soil community. The transformation of a complex system to a simplified system typically leads to lower resource-use efficiency, lower resistance to future disturbance, reduced resilience in terms of the ability to recover from disturbance, and increased disparities in dominance by the remaining organisms.

From an agricultural perspective, loss of soil communities can be seen both positively and negatively. The loss of soil health means a greater reliance on artificial and external inputs of resources to promote plant growth, greater loss of the soil itself due to wind and water erosion, and slowly diminishing productivity. However, if fertilizer and chemical needs are relatively (and artificially) inexpensive compared to the per-acre profit, and the reductionist approach to agriculture continues to provide high crop yields, then the damage to the soil can be ignored without penalty for many years. However, in terms of long-term sustainability, the health of the soil cannot be ignored indefinitely.

reestablish a system of strong and effective checks and balances in even the most damaged, simplified, and chemically dependent agricultural situations. In fact, for some of the cases described in this book, a return to ecosystem-based farming may be the only alternative left for regaining control of agricultural productivity. For example, the ecosystem approach offers hope to the fruit-tree permacultures where green peach aphids are a persistent problem and chemical "solutions" are being rapidly exhausted. Green peach aphids have a large number of natural enemies, including ladybird beetles, lacewings, syrphid flies, parasitic wasps, parasites, and a fungus, *Entomophthora* spp., that attacks aphids.[10] Because the aphids can overwinter in and around crop fields, the establishment of year-round cover vegetation should replace the typical practice of maintaining bare soil.[11] Such cover will create habitat with vertical structure, which will provide resources and a refuge for a diversity of predatory organisms.[12] For example, ladybird beetle adults and larvae are ravenous consumers of aphids, but ladybird beetles require an overwintering refuge, such as large perennial bunchgrasses, from which thousands of adults will emerge in the spring. The availability of overwintering habitat means the predators are residents and do not have to migrate into fields in search of prey and so will be able to reduce prey density long before they become dangerously abundant. If a diversity of predatory species is present, the reliance on a single predator is reduced, as is the probability of a pest species becoming abundant and dominant.

~

Western society, in particular, has worked tirelessly to incorporate technological advances into food and fiber production and to move ever further away from more traditional methods of farming. In doing so, we have also introduced many nonnative, invasive, and pest species while simultaneously reducing the diversity and abundance of the control species, whether native or not. We have created the situation in which we currently find ourselves—trapped in an artificial cycle of chemical dependency with no chance of help from the natural biological assets that we have worked so assiduously to eliminate.

The reestablishment of natural diversity in and around farms appears to be the key to effective pest control. The focus on diversity at every trophic level, including the predators *and competitors,* is in stark contrast to the more modern approach of focusing solely on herbivore species, whether singly or as a group, with the use of chemicals. Every individual species, whether crop or pest, exists and evolved in a natural ecosystem context, and if the context is ignored, the success of the attempts to control pests will be compromised. Thus, pest control, soil health, productivity, community stability, and so on are all dependent on reestablishing an ecosystem context. If a broad spectrum of plant types grows in, around, or near a crop field, a greater diversity of food resources are available to support a greater diversity of herbivorous species, which then provide a diverse support base for a diversity of predators. The diversity within the food chain forms the food web, and the more diverse each level of the food chain, the more stable the food web will be. If each aphid predator requires a different plant on which to overwinter, then greater plant diversity will support greater predator diversity. In this way, if six different predators that prey upon aphids are present, the loss for whatever reason of one of those species of predators will not result in a population explosion of aphids. In short, efforts to maintain trophic diversity within the food web will create a stable, diverse, and protective matrix in which the crop species can be grown with little or no need for chemical intervention. It should go without saying that reduction of chemical use will immediately reduce the threat of new chemically resistant pests and will save the farmer time and money.

Chapter 14

Integrated Systems
and Long-Term Stability

After an entire book describing the processes by which agricultural pest species evolve resistance, it should be clear to readers that the principles of evolutionary biology govern the world's production of food and fiber. Even though many people working in agriculture understand these principles and the consequences of overusing chemicals, it would appear that the implications have not really been taken to heart. For example, the industry's current focus is on creating crops that will attack insects by the internal production of a biochemical; the new strategy is essentially the same as the old one, and the results will be as predictable as ever. Chemical assaults on crop pests, no matter what the origin of the chemical, will stimulate adaptive responses from the pest species. The industry will remain in an evolutionary battle as it tries to counter each adaptive transformation by the pest with newer and "better" technological solutions. The treadmill that is the Red Queen will continue to move and we will continue to run as fast as possible just to stay in one place.

In this chapter, we will explore several examples in which ecological principles are the basis for successfully raising crops, controlling pests, protecting biodiversity, repairing and improving soil, establishing sustainable farming economies, and reducing the costs of food and fiber production. The decisions that led to these changes were neither easy nor taken lightly. They often involved a complete rethinking of farming practices, although in many instances the changes might be viewed as a return to more traditional agriculture, the kind of farming that typified the lives of current farmers' grandparents. In all cases, the farmers involved are convinced that they are practicing "real farming." This is in large measure because they see farming as a working partnership with the soil and the land. The technological advances of the past 60–70 years are welcome if they assist the farmer at his or her job (e.g., air-conditioned cabs on harvesters), but a dependence on those technologies has also created an increasing burden in terms of escalating costs, especially for the many short-term solutions such as chemicals. In the past two decades, the reliance on technology, particularly genetically modified crop plants, has become so complete that many farmers feel they have become divorced from the job itself. That is, they are no longer independent producers, but rather middlemen who apply industrial technology to wrest a product from the soil.

In the larger scope, applying technological solutions to biological problems, and to farming in general, has repercussions that ultimately limit our ability to achieve the objectives we struggle so mightily toward. The slow loss of soil function, of biodiversity, and of ecosystem resilience eats away at the foundation for producing food and fiber. Regardless of the often spectacular, but usually short-term, successes of newer pesticides or pest-resistant crops or of crops that require fewer chemical inputs, ignoring the health of the agro-ecosystem creates a trade-off that we cannot afford. Fortunately for us and our children, ecosystem function appears to be highly resistant and resilient to our insults; it is relatively easy to restore many functions in a remarkably short period of time when the focus turns from *how do we maximize production?* to *how do we protect the source of high productivity?* From some selected examples, we can see that the answer is not necessarily an all-or-nothing approach, but can be achieved with small, yet thoughtful, changes in farming practices.

An Early Voice of Caution

In 1951, as the world of agriculture was beginning to embrace synthetic chemicals, Dr. Reginald Painter at Kansas State University published *Insect Resistance in Crop Plants*, in which he admonished the agricultural world to study and understand the natural attributes of the crop plant itself with regard to insect resistance.[1] At the same time, he warned that insect susceptibility, and therefore resistance, to chemicals was not always genetic. He pointed to numerous examples of studies showing that insect susceptibility to the best insecticides could depend on the types of plants the insects had been raised on. That is, insects feeding on some plants were more susceptible to chemical sprays while those on other plants were less susceptible. Similarly, many studies showed clearly that different varieties of crop plants had different resistances to insect predation. The implications to Painter were twofold: it is necessary to understand the natural ability of plants to protect themselves from predation, and also to understand the influence that plant compounds have on the insects that eat them. He made this remarkable statement (for 1951):

> The evidence presented here indicates that, in a number of cases, control by host-plant resistance as measured by yield compared favorably with control by insecticides. . . . In most problems involving phytophagous insects, the use of insecticides will remain an emergency control measure and, as such, emphatically necessary. It is equally necessary that we attempt to use more permanent control methods that are less costly to growers.[2]

Even today, Painter's insights remain underappreciated and under-investigated. The world of agronomy and crop development did not attempt to manipulate (or even understand) crop genomes for resistance and instead focused on external chemical pest control combined with a tremendous simplification of crop genetic diversity. Had we spent equal time and effort attempting to understand the nature of interactions between crop plants and their herbivores, it is very likely we would not have run headlong toward an agriculture based entirely on growing crops in isolation from their natural environment. Instead, it seems very likely that our agriculture, at the very least, would be based on a polyculture approach that attempted to enhance the resistance of the crops to their pests instead of getting ourselves trapped in a system wherein battling the resistance of the pests is the primary focus.

A Good Start in Australia

Dr. Stephen Powles (University of Western Australia) is one of the foremost authorities on the problems associated with pest resistance to chemicals. His research and publications consistently argue for care and caution: protect the land and protect the farmer's ability to battle unwanted pests. He does not oppose the use of all chemicals but is focused instead on long-term maintenance of farming capability. Powles argues that when we destroy perfectly useful weapons for battling pests, we are not being wise stewards. In fact, Australia now requires specific labeling on herbicides so that farmers can make more-informed decisions about the mode of action of each chemical and can become better educated about the full array of chemical tools available to them.

Australian agriculture provides a prime example of "what can go wrong" with respect to the use of herbicides in farming and ranching. Australians grew ryegrass (*Lolium rigidum*) as a preferred forage species for the sheep (wool) industry and experimented with many genetic varieties to select the best cultivars. Of course, they had a great interest in producing ryegrass that could withstand the chemicals being used to control ryegrass pests. When the global wool market began to undercut Australian wool in the 1980s, many farmers switched to wheat and pasture seed production, in which ryegrass is a persistent weed, and they were faced with an unfortunate situation: they were attempting to control a weed that had been literally designed to resist chemical control. The stage was set. With the combination of vast acreage of ryegrass with high genetic variation and the intensive use of very few, very powerful herbicides, Australian wheat farmers quickly produced ryegrass resistant to as many as seven herbicide modes of action (MOAs), in addition to wild radish (*Raphanus raphanistrum*) resistance to as many as three MOAs.[3] In addition, some of the resistance in ryegrass was a result of "cross-resistance" in which the adaptation to one MOA somehow facilitated the resistance to other herbicides of different MOAs, thereby creating weeds that were essentially ready for the next chemical attack.[4] Thus, even as one herbicide was losing its effectiveness, the next herbicide was already less effective than anticipated.

In 1998, Powles began the Australian Herbicide Resistance Initiative to combat the growing problem, and he has worked closely with other experts and farmers to develop creative solutions to herbicide-resistance issues that were threatening the continued existence of wheat farming in Australia.[5] In particular, the farmers must focus on maintaining a diverse toolkit of effective weapons against weeds. Given the remarkable ability of ryegrass to adapt to herbicides, one important focus is on basic ecology in the form of weed-seed management. That is, control of all weeds should begin with control of the tremendous numbers of seeds they produce. To this end, Stephen Powles and Michael Walsh and their colleagues have developed several mechanical (i.e., nonchemical) approaches to reducing the numbers of viable weed seeds that remain in wheat fields. Their techniques include extra harvest equipment for catching weed seeds, concentrating weed chaff in centralized locations for burning, or mechanical destruction of the seeds. After fine-tuning the techniques, the effect on weed densities in wheat fields has ranged from promising to dramatic. With careful proactive attention to the source of weed seeds, wheat farmers have reduced their reliance on chemicals and the associated costs, and they are slowing the emergence of chemical resistance in ryegrass.

The situation in Australia was and is essentially one of crisis management. The need for a more united effort to protect the wheat industry is having a positive carry-over effect on other farming and ranching concerns. However, these efforts have to be considered as still in their infancy.[6] Other weed species may not be as easy to control using the same mechanical techniques, and longer-term issues regarding loss of soil quality remain. The recognition that herbicides are essentially a dead end as a single avenue for weed control is the most powerful lesson learned. Retooling the industry toward a multifaceted and integrated system of growing wheat (and other crops) will require close attention to the farm field as an agro-ecosystem and to the natural ecosystem surrounding it.

On American Soil

The United States has a well-documented history of soil problems in agricultural systems. The Dust Bowl of the 1930s was the direct result of the adoption of the modern approach toward food and fiber production, as described in chapter 2. Originally, the Midwest of the United States was not viewed as prime agricultural land because the dense root systems of the prairie grasses prevented plows from tilling the soil. With the invention of the steel moldboard plow by John Deere, the technology for cutting through the thick root systems was made available, and the sod could be "busted." Agriculture was revolutionized because the scale of the family farm could be greatly expanded.[7] The subsequent removal of nearly every vestige of perennial prairie vegetation and the sowing of annual crops created a situation in which the soil lay fallow and exposed for months at a time. The combination of exposure, drought conditions, wind and water erosion, the loss of organic material, and the loss of many biotic components resulted in highly degraded agricultural conditions. The Dust Bowl was inevitable and only required the convergence of certain climatic conditions. In many ways, little has changed since then. Perhaps more care is taken to reduce erosion, but the modern era of synthetic fertilizers and pesticides has wrought ever more damage to an already damaged resource.

Soil is a complex, multifaceted living thing that is far more than the sum of the sand, silt, clay, fungi, microbes, nematodes, and other invertebrates. All biotic components interact as an ecosystem within the soil and at the surface, and in relation to the larger components such as herbivores that move across the land. Organisms grow and dig through the soil, aerate it, reorganize it, and add and subtract organic material. Mature soil is structured and layered and, very importantly, it remains in place. Plowing of the soil turns everything upside down. What was hidden from light is exposed. What was kept at a constant temperature is now varying with the day and night and seasons. What cannot tolerate drying conditions at the surface is likely killed. And very sensitive and delicate structures within the soil are disrupted and destroyed.

Conventional tillage disrupts the entire soil ecosystem. Tractors and farm equipment are large and heavy; they compact the soil, which removes air space and water-holding capacity. Wind and water erosion remove the smallest soil particles, which typically hold most of the micronutrients needed by plants. Synthetic fertilizers are added to supplement the loss of soil nutrients but often are relatively toxic to many soil organisms. And chemicals such as pre-emergents, fumigants, herbicides, insecticides, acaricides, fungicides, and defoliants eventually kill all but the most tolerant or resistant soil organisms. It does not take long to reduce a native, living, dynamic soil to a relatively lifeless collection of inorganic particles with little of the natural structure and function of undisturbed soil.

The importance of soil health as the foundation for agricultural health is of almost religious significance for many resource managers and land-care specialists. To them, care of the soil is part of a holistic approach to farming; soil health and sustainable farming are inseparably intertwined and cannot ever be considered in isolation. While the work of many people could be described here, I'll focus on two particularly strong advocates for soil care and protection as examples. The first is Wes Jackson of The Land Institute and the second is Ray Archuleta (and colleagues) at the Natural Resources Conservation Service of the US Department of Agriculture (with Buz Kloot at the Earth Science and Research Institute at the University of South Carolina).

Since 1976, Wes Jackson has advocated a return to a system of farming that is both a more natural fit with the environment and less damaging to the ecosystem services that are essential for sustainable agriculture, especially those provided by the soils.[8] In Jackson's view, modern farming is unsustainable because it defies all ecological rules. When he looks at a typical farm he sees degradation of soil, erosion, loss of diversity, inefficient resource use, genetically weak plants, and the need for regular inputs, as opposed to a natural ecosystem where he sees soil building, no runoff, high diversity with very efficient resource use, high genetic diversity, and no need for fertilizer or pesticides. In a natural system, Jackson sees resiliency, stability, and high productivity in perpetuity; that is, he sees the real rules of farming, all of which are being violated in conventional farming systems.

In the late 1970s, Jackson created The Land Institute in Salina, Kansas, both as a school of farming philosophy and as a demonstration of his vision of farming. Jackson's primary tenet and the research focus of The Land Institute was one that incorporated a more ecological approach to food production: sustainable agriculture must be based on perennial polyculture and the protection of the soil. While the ecological foundation is absolutely sound, there are admitted difficulties. Perennial plants do not behave like annual crop plants, and they invest much of their energy into roots and persistent biomass each season instead of directing it toward the production of seeds that can be harvested for food. And our culture is oriented around foods produced from the crop species we have invested so much time, energy, and research into developing, such as corn and wheat, which are inappropriate for perennial polyculture. Thus, the development of appropriate perennial polycultures that can provide a reliable supply of food for human society requires a considerable amount of planning and development and, of course, demonstration. However, mixtures of perennial species are capable of using resources more fully in comparison to a monoculture; indeed, *overyielding* has been regularly reported.[9] That is, the perennial system produces more per unit area than does a monoculture.

The Land Institute has been a center of research and development of perennial polyculture farming for decades and has demonstrated the possibility of generating high yields from perennial polycultures, particularly with native species, with zero inputs from fertilizers and pesticides.[10] This approach to farming is both a return to a more traditional relationship between the farmer and the land and also a reflection of the understanding that the foundation of agro-ecosystem productivity lies in biodiversity and protection of the resource base, the soil. Nonetheless, despite the recognition of Wes Jackson as an agricultural visionary, widespread acceptance and application of this approach to farming has been slow.

Ray Archuleta's approach to educating farmers about the importance of healthy soil is through application, demonstration, and enthusiasm. The first task is getting farmers to understand the attributes of healthy soil and then to demonstrate the remarkable differences in quality between tillage soils and intact soils. Once the point has been made, farmers are shown how easy it is to heal abused soils. Archuleta's approach is strongly reminiscent of Masanobu Fukuoka's natural farming, and his techniques are often simple ways to remedy the problems and to reverse the damage while teaching the concepts of agro-ecology. As a conservation agronomist at the NRCS East National Technology Center, in Greensboro, North Carolina, Archuleta is not only an advocate for good soil science, but has embraced the power of video for delivering the message as he works with farmers and soil scientists around the United States. This message is delivered in a variety of formats by his equally energetic colleague, soil-scientist-*cum*-videographer Buz Kloot.[11]

The message is basically this: damaged soil cannot provide the services we ask of it, but healthy soil can do more than we ever imagined. Farming practices that damage and reduce soil function and weaken natural interactions with plants have fostered a reliance on fertilizers and pesticides to provide the missing functions. The partnership between plants and soil is mutualistic: plants provide materials that contribute to soil health, and a healthy, complex, functioning soil provides a range of services and resources for the plant. When plants are removed, soils suffer; when soils are damaged, plants suffer.

Archuleta and his colleagues believe, and can demonstrate, that even the most damaged soil can be returned to health in a very short amount of time, even as little as two to three years. The soil protection and restoration practices he espouses may arouse skepticism in farmers trained in modern farming techniques because they contradict current practices, but Archuleta believes his approach is based on sound soil-health principles that improve soil function.

Similarly to Fukuoka's approach, a multi-species cover-crop seed mix is sown during the non-growing season to cover the soil. A diverse range of plants is carefully selected because each species contributes in different ways, such as protecting the soil from wind and water erosion, adding organic material on the surface and in the topsoil, loosening the soil and aerating it, and, very importantly, creating a haven for the proliferation of fungi, bacteria, microorganisms, and macroorganisms, especially during the so-called dormant months. Thus, during part of the year, the farmer attempts to cultivate a natural ecosystem aboveground that will heal the wounds belowground caused by modern agricultural practices. This is accomplished by maintaining a high level of soil activity during the "dormant" months.

In the spring, the cover crop is not killed with herbicides and the soil is not physically disturbed. The vegetation is flattened with a roller to maintain the soil cover, which also protects the habitat of other organisms. Crop seeds are sown with a mechanical seeder through the flattened plants and into the soil. This system allows the soil to continue to heal as a recovering ecosystem, and it protects the soil surface from evaporation, heat, and the direct impact of rain.

Heavy farm equipment is not driven onto the fields except as needed for seeding, harvesting, and flattening of the cover crop, and the soil is no longer disturbed with any kind of plowing or tillage. Without regular compaction, rapid recovery of soil structure is favored: soil organisms improve the soil's pore space, which greatly increases infiltration of rainwater and reduces puddling and erosion from runoff, and organic material is quickly incorporated into the topsoil by the growing community of invertebrates such as earthworms.

Farmers adopting this system of soil management from North Carolina to North Dakota have reported rapid and very positive gains. The use of legumes and other species in the cover crop adds a range of nutrients to the soil and greatly reduces and even eliminates the need for synthetic fertilizers. As the microbial and invertebrate communities grow, the nutrients in the vegetation are returned to the soil as the plants decompose. Protecting the soil surface reduces evaporation and increases rainwater infiltration, which reduces irrigation needs. The invertebrate community rapidly develops a trophic structure with numerous species of predators, and the need for insecticides is greatly reduced. Likewise, without exposed soil, weeds are suppressed and fewer herbicides are needed. This also reduces the need for operating farm equipment and thus lessens the associated costs. Most importantly, yields are as high as with conventional tillage methods. In some cases, farmers have eliminated the need for pesticides and fertilizers altogether.

The sustainable approaches to agro-ecosystems described here and in the previous chapter make the maintenance or reestablishment of healthy soils a core principle. And what could make more sense, since all crop plants absorb water and nutrients through their roots? Wes Jackson advocates a system of low-input agriculture based on perennial polycultures, and he has demonstrated the viability of such an approach. Indeed, a society based on the production of food from perennial plants grown in polycultures makes tremendous ecological sense. However, the vast majority of agriculture is based on annual monocultures. Is there a way to protect and enhance soil function in annual crop systems? Ray Archuleta and his colleagues (and Fukuoka before them) are demonstrating that creative modifications to the way we produce an annual crop are viable if we think in broader terms about what "agriculture" looks like. Most importantly for the future of "modern" agriculture, the reestablishment of biological controls and of healthy soil function can support crop monoculture systems, and this recovery can take place in even the most abused and biologically "dead" soils.

Putting Out Fires in California

California is an agricultural region unlike any other. It is the most diverse and productive in the world, with year-round farms in the southern Imperial Valley, tree crops in the eastern Central Valley, annual and biennial crops throughout the state, and the wine vineyards of the central and north coast. The coastal areas are also home to cool-season crops such as lettuce, broccoli, and garlic. The inland areas grow large acreages of cotton, wheat, rice, almonds, grapes, sugar beets, tomatoes, walnuts, pomegranates, oranges, peaches, and many more crops. Most crop species and their mixtures are centered in particular climate regions of the state, but the fantastic diversity of crops has also attracted a fantastic diversity of crop pests. And because of the tremendous importance of agriculture to the California economy, which is among the top ten largest *national* economies in the world, threats to the health of agriculture are taken very seriously.

The University of California system has taken the lead in the battle to protect agriculture in California. The immense task requires innovation and creativity, and it demands success because the consequences of failure are also immense. For decades, resource managers and extension specialists at most universities espoused conventional farming practices that included the use of pesticides, fertilizers, clean farming techniques, deep tilling, and many other practices that were tremendously damaging to the agro-ecosystems and the soils. Decisions were driven by the most up-to-date research, almost all of which came from universities themselves, and represented cutting-edge knowledge within a world that had accepted new technologies as the best solutions for the goal of increasing farm productivity. In many ways, the treadmill of the Red Queen was powered by the technological advancements that flowed from universities (which also provided the highly skilled workforce for the agrochemical industry). Likewise, it is at the universities where the rethinking and retooling process is now under way, leading us to solutions for working with nature instead of against it.

The field of integrated pest management (IPM) emerged in the 1990s as a solution to the ever-growing threat of crop pests, and, although it is a logical extension of the conventional approach to food and fiber production, it represented a repositioning within university extension services. IPM has become the standard approach for modern agriculture in many regions and for particular crops. At the University of California, Davis, a research university in the middle of the Central Valley of California, the management issues concerning the accelerating numbers of invasive, resistant, and potentially troublesome species in the region demand constant attention and research. The school is a nexus for scientists, graduate students, landowners, and corporations focused on California agriculture. UC Davis has developed detailed IPM pest-management guidelines for several major economic crops with this as the underlying philosophy:

> Rather than simply eliminating the pests you see right now, using IPM means you'll look at environmental factors that affect the pest and its ability to thrive. Armed with this information, you can create conditions that are unfavorable for the pest.[12]

The UC Davis IPM handbook for cotton is very reminiscent of a farmer's almanac in the sense that all aspects of the farmland should be considered throughout the many steps involved in growing cotton.[13] Farmers are reminded to consider the history of the land, the surrounding crops and their pests, and the use of secondary crops to mitigate cotton pests. Farmers should also choose the cotton cultivar with regard to the local soil conditions and the known pest issues. Planting should be timed according to weather patterns in order to ensure greater success. Most importantly, the IPM approach reminds farmers to be cognizant that IPM

> makes use of all available control strategies, including cultural, host plant resistance, biological, and chemical controls to manage pests. Natural enemies are an extremely important component of integrated pest management of cotton insects and mites.[14]

Thus, modern farming practices, in the age of chemical pest resistance, now recognize the need for integrated multifaceted strategies that include the properties of the surrounding ecosystem.

The 100-page IPM book for cotton begins with interesting advice: choose the field based on the history of the location, the neighboring crops, and the pest history of the field, and choose the best variety of cotton for the soil conditions and the local pest species. In other words, consider the agro-ecosystem and its attributes rather than just planting the crop and then waiting for and dealing with problems. Once a crop is planted, the farmer is advised to monitor the crop in regard to six specific plant-developmental stages and to monitor the pests present at those stages. Chemicals are to be used only at certain plant stages when pests are most vulnerable and when potential pest damage to the crop is highest. For example, a leaf has whiteflies (*Bemisia tabaci*) if three or more adults are found, but chemical application isn't advised until 40 percent of the leaves are infested. Similarly, treatment for lygus bugs (*Lygus hesperus*) is not necessary from 10 days after flowering because they do not reduce yield from that point. Farmers are advised to survey and sample for pests and to keep detailed records of their findings.

When chemicals are used, care must be taken to reduce the impact on natural pollinators and pest enemies.

Natural enemies are an extremely important component of integrated pest management of cotton insects and mites. Common natural enemies include lacewings, bigeyed bugs, damsel bugs, minute pirate bugs, lady beetles, thrips, and several parasitic wasps. Lacewings, lady beetles, and parasitic wasps help control cotton aphids. Spider mite populations can be controlled by predatory mites and thrips. Lepidopterous larvae can be controlled or suppressed by several species of natural enemies. Research has shown that 99% of beet armyworm, cabbage looper, and cotton bollworm eggs and early-instar larvae are consumed by predators *in fields that have a natural population of predators and parasites.* Some insects, such as thrips, can be predators (feeding on spider mite eggs) as well as plant feeders. Generally, the beneficial aspects of thrips outweigh the damage to seedling cotton.[15] [Italics added.]

It is very clear that control of crop pests is no longer solely the domain of chemical solutions, and this understanding has emerged from the recognition that chemicals represent short-term solutions at best. The list of 55 chemicals in the IPM guidelines for cotton gives information about the specificity for certain pests, persistence with regard to pest species, and persistence with regard to the natural enemies of the pest species. The cautionary approach of considering all of the options and using multiple techniques that are timed for greatest effect for controlling pest problems is the hallmark of IPM.

∽

The examples presented here are just a glimpse into some of the simple and sometimes creative ways for repairing and redirecting the current agricultural system of food and fiber production. The many examples of high yields from more diverse systems coupled with improved ecosystem services and long-term sustainability suggest that there are viable alternatives to the current approach to agriculture. Unfortunately, the industries supporting the modern system of agriculture are tremendously invested in the path of pesticides, fertilizers, and genetically modified and simplified crop cultivars. Overcoming the inertia inherent in the culture of growing food and fiber crops will be a tremendous undertaking. However, the broader question, and one that must be addressed, is whether the current system with its ever-growing problems is one that can be sustained indefinitely into the future. Given the documented declines in soil fertility and losses to erosion across the United States and the world, and increasing numbers of pests with increasing levels of pesticide resistance, it seems that we have little choice but to think very hard about *how* we go about the business of producing food. And as we do so, we must remember that the best and most successful path forward will be one that incorporates a clear understanding of the rules of evolutionary biology.

Chapter 15

Epilogue: Putting All of Our Eggs in a Diversity of Baskets

The chemical treadmill, and the future it portends, is very well understood by many in the agricultural world, but the scale and momentum of modern agriculture are enormous. Nonetheless, armed with knowledge of both the current problems and their origins, and with viable solutions, why have we made so little progress? There are many factors that prevent rapid change, but the overwhelming factor is *cultural inertia*—agriculture is an economic juggernaut, and the practices and technology that support our use of agricultural chemicals have become entrenched over the past few decades. To change the way we do things now will require a nearly complete philosophical reversal of our approach to growing food and fiber plants. But do we have a choice? If failure to act is worse in the long run than the short-term discomfort of adopting a new way of thinking, then we do not. However, achieving a major shift toward an ecosystem-based approach to farming will require a consensus among the all important elements of the farming world—farmers, seed producers, agrochemical producers, the biotechnology industry, farm equipment manufacturers, state and federal governments, and consumers.

With 7 billion people to feed and a growing demand for farm products in so many economic sectors, we cannot afford to ignore technology's benefits. It may not be possible to return to the farming practices that prevailed in the days before modern pesticides. What is necessary is to farm more intelligently. We knew the principles of evolutionary biology in 1950, and we knew why resistance to DDT was occurring around the world, but our mistakes over the subsequent decades resulted from our failure to appreciate the power of nature to treat any action of humans as no more than another environmental stress. Now we have no excuses; we cannot exempt ourselves from the rules of the game, we cannot find a way to bend those rules in our favor. What we *can* do is to play the game as intelligent creatures and to anticipate events and consequences. What we *can* do is stay ahead rather than always playing catch-up. The Red Queen is not a punitive system of management, but a set of rules that describe a game for an unlimited number of players, and one that produces survivors, not winners.

After 60 years, the promise of chemical solutions to pest problems on farms has not been realized, nor has biotechnology in the past 20 years made any greater inroads. A shift back toward an ecosystem approach to farming will not necessarily be an "all or nothing" event, and it could be daunting for large-scale commercial farming operations. However, the current research on no-till farming and cover-crop innovations demonstrates the ability of farmers to achieve high yields relatively quickly with very low inputs. The return of soil health and more biodiverse agro-ecosystems will offer benefits in terms of reduced costs, more stable yields, and better soil stability. A focus on sustainable agro-ecology will favor more stable, more resistant, and more resilient crops and agricultural practices. Most importantly, an integrated approach to pest control will interrupt the Red Queen cycle and greatly reduce chemical dependency and the associated costs.

It is important to recognize that stepping off the chemical treadmill will depend on reestablishing ecosystem diversity and function. The unfortunate state of modern agriculture came about because we relied on simple answers to very complex questions; we treated pests as an isolated problem when they are actually embedded in a matrix of interconnected variables that prevent simple solutions from being effective. Western culture has a very long history of attempting to understand a phenomenon by extracting it from its natural context and then treating it as if it were disconnected from the rest of the world. This approach is often characteristic of Western science, culture, and thinking, and it engenders an anthropocentric approach toward other species. That is to say, for the most part we as humans tend to think of ourselves as exempt from natural rules and controls, and we apply that simplicity to the systems that we are attempting to manipulate. Why do we do this?

Science approaches new discoveries using an *inductive reasoning* approach: we collect information and use that information to create a theoretical framework that is used to explain any new information. From there, more evidence is added to the framework to clarify the theory, further allowing us to explain new information. Think of a jigsaw puzzle: as the pieces are gradually added to the puzzle, the picture begins to emerge and our understanding increases. The more pieces that are added, the clearer the picture becomes. It is not necessary to complete the puzzle to have a relatively clear understanding of the picture. However, once the human mind has a mental picture of the way something works, it tends to force all new information through that same cognitive filter. That is, we shift to *deductive reasoning*: we have an understanding of how something works and we attempt to explain new information in terms of that understanding. With reference to the jigsaw puzzle, we believe that every new piece of information should fit somewhere, and we find it impossible to believe that a piece might belong to a different puzzle altogether. Eventually, we accumulate enough information that cannot be explained using the current theory, and it becomes necessary to reconfigure our thinking. The great philosopher Karl Popper argued that the willingness to acknowledge alternative explanations (i.e., "falsifiability") is a hallmark of real and objective science.[1] Scientists must always accept the risk of refutation while in the pursuit of new knowledge. Such a change in perspective, what Thomas Kuhn called a "paradigm shift," is often very difficult and often staunchly resisted.[2] Nonetheless, when the flaws inherent in a particular worldview become not just evident but overwhelming, a change in perspective must precede realistic progress.

Thus, we are in need of a paradigm shift in the way we perceive our relationship with the land, with nature, and with the process of agriculture. Such a shift in the way we see and interpret natural phenomena will be difficult for several reasons. First, it requires breaking free of the approaches we have traditionally been taught, and this means *thinking* about the world in a new and different way. Second, a long-standing worldview is always associated with very strong and established infrastructures; adopting a new worldview means disassembling the existing infrastructures and creating new ones. These necessarily include scientific, social, cultural, and economic institutions. Third, adopting a new worldview means convincing others that their long-held beliefs might be mistaken and that there is a need to change. It is certainly not surprising that reformers encounter rigid resistance, regardless of the evidence in favor of change. In fact, arguments against a new worldview can sound quite convincing and logical because they are based on existing and commonly shared understandings. Such arguments are very difficult to overcome.[3] A paradigm shift in science is similar to a cultural revolution; it can be slow, difficult, and painful. And yet for agriculture, it is absolutely necessary if there is to be any hope of a sustainable future.

NOTES

Preface

1. Lewis Carroll, *Through the Looking-Glass, and What Alice Found There* (London: MacMillan, 1871). See the quotation at the beginning of the book.

Chapter 1

1. Darwinian fitness is a measure of how well an individual succeeds at passing its genes on to the next generation. The "most fit" individuals are those who are able to survive to reproduction and give rise to more offspring than the other individuals in the population.
2. I will use the terms *population* and *species* interchangeably. In evolutionary biology and ecology, the focus is on populations in particular habitats rather than the entire species, which can be made up of numerous populations living in very different habitats. Populations adapt to their particular conditions, but very rarely does an entire species adapt simultaneously.
3. Of course, this assumes the genetic variation necessary for adapting to that stress is present.
4. See: S. J. Gould, "The Misnamed, Mistreated, and Misunderstood Irish Elk," in *Ever Since Darwin* (New York: W. W. Norton, 1977), 79–90.
5. These extinctions are described in: Peter Matthiessen, *Wildlife in America* (New York: Viking Press, 1959).
6. The small vs. large comparison in animals is equivalent to a short-lived annual vs. long-lived perennial species comparison in plants. The large majority of problematic weeds in the world are annual species that grow quickly and produce large numbers of seeds. See: T. S. Prather, J. M. DiTomaso, and J. S. Holt, "Herbicide Resistance:

Definition and Management Strategies," University of California, Division of Agriculture and Natural Resources Publication 8012, 2000.

Chapter 2

1. Lynn White, "The Historical Roots of Our Ecologic Crisis," *Science* 155 (1967): 1203–7.
2. *Monoculture* is the production of a single crop in a field, as opposed to *polycultures* and *intercropping*, where multiple crops are grown simultaneously in a field.
3. Reviewed in M. A. Altieri, "Linking Ecologists and Traditional Farmers in the Search for Sustainable Agriculture," *Frontiers in Ecology and the Environment* 2 (2004): 35–42.
4. Ibid.
5. Reviewed in: E. C. Oerke, "Crop Losses to Pests," *Journal of Agricultural Science* 144 (2006): 31–43.
6. J. Fernandez-Cornejo, C. Hallahan, R. Nehring, and S. Wechsler, "Conservation Tillage, Herbicide Use, and Genetically Engineered Crops in the United States: The Case of Soybeans," *AgBioForum* 15, no. 3 (2012): 231–41.
7. Some would argue that, in fact, these chemical weapons continue to *destroy* the soil, as will be discussed in chapters 13 and 14. See also: S. O. Duke and S. B. Powles, "Glyphosate-Resistant Crops and Weeds: Now and in the Future," *AgBioForum* 12, nos. 3 and 4 (2009): 346–57.
8. I. R. Kennedy, F. Sanchez-Bayo, and R. A. Caldwell, eds., "Cotton Pesticides in Perspective," Australian Cotton Cooperative Research Centre, 2000.
9. S. O. Duke, "Why Have No New Herbicide Modes of Action Appeared in Recent Years?," *Pest Management Science* 68 (2012): 505–12.
10. See: US Environmental Protection Agency, www.epa.gov/opp 00001/pestsales/07pestsales/sales2007.htm#2_1. These data are the most recent available from the US government.
11. Food and Agriculture Organization of the United Nations (UNFAO), http://faostat3.fao.org/home/index.html.

12. First estimate from UNFAO; second estimate from USDA Economic Research Service, www.ers.usda.gov/data-products /agricultural-productivity-in-the-us.aspx#28247.

13. R. Hoppe, J. MacDonald, and P. Korb, "Small Farms in the United States: Persistence Under Pressure," Economic Information Bulletin No. EIB-63, 2010.

14. It is eye-opening to explain an equivalent situation in which a group of ten students work on a project, but nine of them do 25 percent of the work and one student does 75 percent of the work. Such economics can be appreciated by anyone.

15. Hoppe et al., 2010.

16. Ibid. This survey notes that 76 percent of the 204,000 largest farms received government payments for commodity crops, while only about 28 percent of the 1,646,000 smallest farms received such assistance.

17. Ibid.

18. "Soil doesn't need to rest; soil needs to be busy all the time." (Buz Kloot, Earth Sciences and Resources Institute, University of South Carolina, from a seminar given at the University of South Carolina Aiken, October 26, 2012.)

19. See, for example: P. C. Hoffman, R. D. Shaver, and D. A. Undersander, "Utilizing Corn Stalk Residues for Dairy Cattle," www.uwex .edu/ces/dairynutrition/documents/UtilizingCornStalkResiduesfor DairyCowsandHeifersv3.0.pdf; R. Jarabo et al., "Cornstalk from Agricultural Residue Used as Reinforcement Fiber in Fiber-Cement Production," *Industrial Crops and Products* 43 (2013): 832–39; N. C. Moore and K. McAlpine, "Microbial Team Turns Cornstalks and Leaves into Better Biofuel," *Michigan News* (University of Michigan), August 19, 2013.

20. See, for example: Department of Crop and Soil Sciences, Michigan State University, "Soil Ecology and Management," 2004, www .safs.msu.edu/soilecology/soilbiology.htm.

21. M. Fukuoka, *The One-Straw Revolution* (Emmaus, PA: Rodale Press, 1978). This influential book provides a discussion of soil health as the foundation of sustainable agriculture, and how easily soil can be damaged, but also restored.

22. Michael Pollan, *In Defense of Food* (New York: Penguin Press, 2008).
23. Hoppe et al., 2010.
24. Ibid.
25. R. M. May and A. P. Dobson, "Population Dynamics and the Rate of Evolution of Pesticide Resistance," in *Pesticide Resistance: Strategies and Tactics for Management* (Washington, DC: National Academy Press, 1986), 170–93. The authors provide US data that have been widely cited elsewhere for insect losses of 7 percent in the 1940s and 13 percent in the 1980s and 1990s. However, their estimate for losses including diseases and weeds was 32 percent, and the authors suggest this loss has not changed since medieval times.
26. Oerke, 2006. Data for 2001–2002 from CABI Crop Protection Compendium for worldwide production of wheat, rice, maize, potatoes, soybeans, and cotton. The author calculates that losses *without* crop protection would average 69 percent. See also: D. Pimentel. "Environmental and Economic Costs of the Application of Pesticides Primarily in the United States," *Environment, Development, and Sustainability* 7 (2005): 229–52.
27. P. A. Matson, W. J. Parton, A. G. Power, and M. J. Swift, "Agricultural Intensification and Ecosystem Properties," *Science* 277 (1997): 504–9.

Chapter 3

1. Phillips McDougall, Crop Protection and Biotechnology Consultants (company website), www.phillipsmcdougall.com/.
2. Many plants and some animals can reproduce asexually or via parthenogenesis, which results in new individuals that are genetically identical. In theory, as the degree of habitat unpredictability increases, the value of such reproductive strategies is decreased.
3. See the example of bacteria and houseflies, chapter 1.
4. This concept is called Liebig's Law of the Minimum—the idea that every species relies on a large number of resources in the environment, but at least one resource will always be in short supply and will limit the growth of the population. For example, a particular nutrient could be uncommon. If that limitation is alleviated, one of the other necessary resources will become the limiting factor.

Chapter 4

1.   P. A. Matson et al., "Agricultural Intensification and Ecosystem Properties," *Science* 277 (1997): 504–9.

2.   R. L. Nielsen, "Historical Corn Grain Yields for Indiana and the U.S.," Corny News Network, Agronomy Department, Purdue University, August 2012, www.agry.purdue.edu/ext/corn/news/timeless /YieldTrends.html, figure 1.

3.   University of Illinois, Department of Agricultural and Consumer Economics, "Farmdoc," 2014, www.farmdoc.illinois.edu/.

4.   Mark Bertness and Ragan Callaway, "Positive Interactions in Communities," *Trends in Ecology and Evolution* 9 (1994): 191–93.

5.   There are a number of publications concerning the loss of and the importance of agro-ecosystem complexity for sustainability, pest resistance, and essential ecosystem services. Examples include: G. W. Cox and M. D. Atkins, *Agricultural Ecology* (San Francisco: Freeman, 1979); E. Schultze and H. A. Mooney, eds., *Biodiversity and Ecosystem Function* (New York: Springer, 1993); and M. A. Altieri, *Biodiversity and Pest Management in Agroecosystems* (New York: Haworth Press, 1994).

6.   Alfred Korzybski, *Science and Sanity* (New York: International Non-Aristotelian Library Publishing Co., 1933).

7.   Ghazi Al-Karaki, B. McMichael, and John Zak, "Field Response of Wheat to Arbuscular Mycorrhizal Fungi and Drought Stress," *Mycorrhiza* 14 (2004): 263–69.

8.   P. S. Ward, "A Synoptic Review of the Ants of California (Hymenoptera: Formicidae)," *Zootaxa* 936 (2005): 1–68.

9.   Robert T. Paine, "Food Web Complexity and Species Diversity," *American Naturalist* 100 (1966): 65–75; Robert T. Paine, "A Note on Trophic Complexity and Community Stability," *American Naturalist* 103 (1969): 91–93.

10.   Richard A. Minnich, "Fire Mosaics in Southern California and Northern Baja California," *Science* 219 (1983): 1287–94. This paper provides a discussion of the creation of habitat mosaics from wildfire, particularly as it applies to areas such as northern Mexico, where fire suppression is largely unmanaged.

11.   M. A. Altieri, "The Ecological Role of Biodiversity in Agroecosystems," *Agriculture, Ecosystems, and Environment* 74 (1999): 19–31.

12.   M. Fukuoka, *The One-Straw Revolution* (Emmaus, PA: Rodale Press, 1978).

13.   See box 2-1 concerning the *Cactoblastis* moth in Australia; see also: R. H. MacArthur and E. O. Wilson, *The Theory of Island Biogeography* (Princeton, NJ: Princeton University Press, 1967).

14.   Jesse H. Ausubel, Perrin S. Meyer, and Iddo K. Wernick, "Death and the Human Environment: The United States in the 20th Century," *Technology in Society* 23, no. 2 (2001): 131–46.

15.   In fact, much of modern medicine was based on antigenic responses in which the human immune system is essentially prepared for the appearance of the pathogen. This pre-infection strengthening of the immune system prevents the pathogen from ever gaining a foothold, and so the population explosion of the pathogen never takes place. If anything, modern agro-science has taken the exact opposite approach by systematically reducing the ability of crop species to defend themselves (as discussed in box 4-1, "Secondary compounds," and also the section on Reginald Painter in chapter 14).

16.   This would not be true of general medications such as multipurpose, broad-spectrum antibiotics.

Chapter 5

1.   R. N. Mack and M. Erneberg, "The United States Naturalized Flora: Largely the Product of Deliberate Introductions," *Annals of the Missouri Botanical Garden* 89 (2002): 176–89.

2.   G. W. Hendry and M. P. Kelly, "The Plant Content of Adobe Bricks," *California Historical Society Quarterly* 4 (1925): 361–73; G. W. Hendry, "The Adobe Brick as an Historical Source," *Agricultural History* 5 (1931): 110–27.

3.   United States Department of Agriculture, "State Noxious-Weed Seed Requirements Recognized in the Administration of the Federal Seed Act (2014),"

4.   Australian Government, Department of Agriculture, "Seed Contaminants and Tolerance Tables," 2014, www.daff.gov.au/biosecurity /import/plants-grains-hort/contaminants-tolerance.

5.   R. N. Mack, "Cultivation Fosters Plant Naturalization by Reducing Environmental Stochasticity, *Biological Invasions* 2 (2000): 111–22.

6. F. Gould, "The Evolutionary Potential of Crop Pests," *American Scientist* 79 (1991): 496–507.

7. *Ploidy* refers to the number of copies of chromosomes. All animals and many plant species have two copies of each chromosome and are diploid (2N). However, depending on their evolutionary history, plants may be triploid (3N), tetraploid (4N), or even octoploid (8N). Any number is possible, but more than two is considered *polyploid* and is a characteristic of the majority of plant species.

8. M. Scudellari, "Genomes Gone Wild," *Scientist*, January 1, 2014.

Chapter 6

1. Phillips McDougall, Vineyard Business Centre, "R & D Trends in Crop Protection," presentation to the Annual Biocontrol Industry Meeting, 2012, www.abim.ch/fileadmin/documents-abim /Presentations_2012/ABIM_2012_6_McDougall_John.pdf; S. O. Duke, "Why Have No New Herbicide Modes of Action Appeared in Recent Years?" *Pest Management Science* 68 (2012): 505–12.

2. Weed Science Society of America and Herbicide Resistance Action Committee, "Summary of Herbicide Mechanism of Action," 2014, http://wssa.net/wp-content/uploads/HerbicideMOAClassification .pdf.

3. Duke, 505–12. See also: J. K. Norsworthy et al., "Reducing the Risks of Herbicide Resistance: Best Management Practices and Recommendations," *Weed Science* 60 (2012): 31–62.

4. Norsworthy et al., 31–62. This paper reviews the issue and makes a number of recommendations for adopting practices for more effective chemical use and for conserving the value of important chemical MOAs. See also: Stephen O. Duke and Stephen B. Powles, "Glyphosate: A Once-in-a-Century Herbicide," *Plant Management Science* 64 (2008): 319–25. This paper reviews the importance of glyphosate to agriculture worldwide.

5. W. K. Vencill et al., "Herbicide Resistance: Toward an Understanding of Resistance Development and the Impact of Herbicide-Resistant Crops," *Weed Science* 60 (2012): 2–30.

6. Duke and Powles, 319–25.

7. A. P. Appleby of Oregon State University has published an "Herbicide Company 'Genealogy'" that tracks the growth in size and re-

duction in number of herbicide companies in the United States; see: http://cropandsoil.oregonstate.edu/system/files/u1473/tree.pdf.

8. Weed Science Society of America, 3.

9. These examples are reported in Norsworthy et al. (2012) with original references.

10. H. J. Beckie, "Herbicide-Resistant Weeds: Management Tactics and Practices," *Weed Technology* 20 (2006): 793–814.

11. I. Heap, "The International Survey of Herbicide Resistant Weeds," August 1, 2014, www.weedscience.com/summary/home.aspx.

12. J. Mallet, "The Evolution of Insecticide Resistance: Have the Insects Won?" *Trends in Ecology and Evolution* 4 (1989): 336–40.

13. See: Insecticide Resistance Action Committee International MoA Working Group, 2012, www.irac-online.org/.

14. Michigan State University, "Arthropod Pesticide Resistance Database," August 1, 2014, www.pesticideresistance.com/search.php.

15. J. E. Casida and G. B. Quistad, "Golden Age of Insecticide Research: Past, Present, or Future?" *Annual Review of Entomology* 43 (1998): 1–16.

16. Ibid.

17. C. Wilson and C. Tisdell, "Why Farmers Continue to Use Pesticides Despite Environmental, Health and Sustainability Costs," *Ecological Economics* 39 (2001): 449–62.

18. D. Pimentel, "Environmental and Economic Costs of the Application of Pesticides Primarily in the United States," *Environment, Development, and Sustainability* 7 (2005): 229–52.

Chapter 7

1. T. H. Suchanek et al., "Evaluating and Managing a Multiply Stressed Ecosystem at Clear Lake, California: A Holistic Ecosystem Approach," pp. 1233–65 in *Managing for Healthy Ecosystems: Case Studies*, ed. W. L. Lasley et al. (Boca Raton, FL: CRC/Lewis Press, 2002).

2. Ibid.

3. P. B. Moyle, *The Inland Fishes of California*, revised and expanded edition (Berkeley, CA: University of California Press, 2002).

4. R. Carson, *Silent Spring* (New York: Houghton Mifflin, 1962).

5. A. X. Silva et al., "Insecticide Resistance Mechanisms in the Green Peach Aphid *Myzus persicae* (Hemiptera: Aphididae) I: A Transcriptomic Survey," *PLoS ONE* 7, no. 6 (2012): e36366, doi: 10.1371/journal.pone.0036366.

6. J. L. Capinera, "Featured Creatures: Green Peach Aphid," University of Florida Pub no. EENY-222 (revised 2005), http://entnemdept.ufl.edu/creatures/veg/aphid/green_peach_aphid.htm.

7. Silva et al.

8. Capinera.

9. Arthropod Pest Resistance Database, Michigan State University .www.pesticideresistance.com/display.php?page=species&arId=384, accessed August 2, 2014.

10. The list of banned and restricted pesticides is published regularly as part of the Rotterdam (www.pic.int/) and Stockholm Conventions (http://chm.pops.int/Home/tabid/2121/mctl/ViewDetails/EventModID/871/EventID/407/xmid/6921/Default.aspx).

Chapter 8

1. See: National Cotton Council of America, 2013, www.cotton.org/.

2. Ibid.

3. Environmental Justice Foundation, *The Deadly Chemicals in Cotton* (London, UK: EJF in collaboration with Pesticide Action Network UK, 2007), ISBN no. 1-904523-10-2.

4. K. R. Kranthi et al., "Insecticide Resistance in Five Major Insect Pests of Cotton in India," *Crop Protection* 21 (2002): 449–60.

5. United States Department of Agriculture, "Agricultural Chemical Usage—2010 Corn, Upland Cotton, and Fall Potatoes," May 25, 2011, http://www.nass.usda.gov/Surveys/Guide_to_NASS_Surveys/Chemical_Use/index.asp#description.

6. *Anthonomus* is a true beetle (coleoptera) and *Helicoverpa* and *Heliothis* are noctuid moths. The moths are in a subfamily, heliothinae, in which several important species had adapted to all widely used agrochemicals around the world as of 1998. See: A. R. McCaffery, "Resistance to Insecticides in Heliothine Lepidoptera: A Global View," *Philosophical Transactions of the Royal Society of London*, Series B 353 (1998): 1735–50.

7.   National Research Council (NRC), *Ecologically Based Pest Management* (Washington, DC: National Academies Press, 1996).

8.   D. G. Botrell and P. L. Adkisson, "Cotton Insect Pest Management," *Annual Review of Entomology* 22 (1977): 451–81.

9.   P. L. Adkisson and S. J. Nemec, "Comparative Effectiveness of Certain Insecticides for Killing Bollworms and Tobacco Budworms," Texas Agricultural Experiment Station Bulletin 1048, 1966.

10. United Nations Conference on Trade and Development (UNCTAD), "Cotton Pests and Diseases," 2011, http://www.unctad .info/en/Infocomm/Agricultural_Products/Cotton/Crop/Cotton -pests-and-diseases/.

11. Botrell and Adkisson.

12. University of California, "UC IPM Pest Management Guidelines: Cotton," Publication 3444, 2013, www.ipm.ucdavis.edu/PMG /selectnewpest.cotton.html.

13. See, for example: W. K. Vencill et al., "Herbicide Resistance: Toward an Understanding of Resistance Development and the Impact of Herbicide-Resistant Crops," *Weed Science* 60 (2012): 2–30; J. K. Norsworthy et al., "Reducing the Risks of Herbicide Resistance: Best Management Practices and Recommendations," *Weed Science* 60 (2012): 31–62; University of California, "UC IPM"; National Research Council (NRC); Botrell and Adkisson.

14. NRC.

15. P. A. Matson et al., "Agricultural Intensification and Ecosystem Properties," *Science* 277 (1997): 504–9.

16. UNCTAD.

17. Kranthi et al.

18. USDA-NASS data retrieved from Mississippi State University, Department of Biochemistry, Molecular Biology, Entomology, and Plant Pathology, "Cotton Crop Loss Data," 2013, www.entomology .msstate.edu/resources/cottoncrop.asp.

19. Botrell and Adkisson.

Chapter 9

1.   The common name for maize in the United States is *corn*, which literally means "seed." In this chapter, I will refer to maize as "corn," but I will use the term *maize* in subsequent chapters.

2. US Department of Agriculture, Economic Research Service, "Corn," USDA ERS website, 2013, www.ers.usda.gov/topics/crops/corn /background.aspx#.Ufpb8qwqj90.

3. Corn Refiners Association, "2013 Corn Annual," www.corn .org/; M. C. Lott, "The·U.S. Now Uses More Corn for Fuel than for Feed," *Scientific American* (online), October 2011, http://blogs .scientificamerican.com/plugged-in/2011/10/07/the-u-s-now -uses-more-corn-for-fuel-than-for-feed/; USDA Economic Research Service.

4. W. L. Brown et al., "Origin, Adaptation, and Types of Corn," *National Corn Handbook* (Ames, IA: Iowa State University, Cooperative Extension Service, 1985).

5. H. C. Chiang, "Pest Management in Corn," *Annual Review of Entomology* 23 (1978): 101–23.

6. Ibid.

7. University of California, "UC IPM Pest Management Guidelines: Corn," 2011, www.ipm.ucdavis.edu/PDF/PMG/pmgcorn.pdf.

8. K. L. Flanders et al., "Maize Insects in North America," in *Radcliffe's IPM World Textbook* (Minneapolis, MN: University of Minnesota, 2006), http://ipmworld.umn.edu/chapters/maize.htm #Spider; University of California, "UC IPM Pest Management Guidelines: Corn"; C. James, "Global Review of Commercialized Transgenic Crops: 2002 / Feature: Bt Maize," International Service for the Acquisition of Agri-Biotech Applications (ISAAA) Briefs no. 29, 2003.

9. The "Old Rotation" research project on crop rotations and soil quality was begun at Auburn University about 1896; see: www.ag .auburn.edu/agrn/cotton.htm. More on this topic will be presented in chapter 13.

10. A. E. Sammons et al., "Behavioral and Feeding Assays Reveal a Western Corn Rootworm (Coleoptera: Chrysomelidae) Variant That Is Attracted to Soybean," *Environmental Entomology* 26 (1997): 1336–42.

11. M. J. Curzi et al., "Abnormally High Digestive Enzyme Activity and Gene Expression Explain the Contemporary Evolution of a *Diabrotica* Biotype Able to Feed on Soybeans," *Ecology and Evolution* 2 (2012): 2005–17.

12. C. C. Chu et al., "Gut Bacteria Facilitate Adaptation to Crop Rotation in the Western Corn Rootworm," *Proceedings of the National Academy of Sciences USA* 110 (2013): 11917–22.

Chapter 10

1. University of California, "UC IPM Pest Management Guidelines: Western Grapeleaf Skeletonizer," 2008, www.ipm.ucdavis.edu /PMG/r302301011.html.

2. US Department of Agriculture, Economic Research Service, "Adoption of Genetically Engineered Crops in the U.S.," USDA ERS website, 2013, www.ers.usda.gov/data-products/adoption-of -genetically-engineered-crops-in-the-us.aspx#.Uf_4xawqj90.

3. GMO Compass, "Genetically Modified Plants: Global Cultivation on 134 million hectares," 2009, www.gmo-compass.org/eng/agri _biotechnology/gmo_planting/257.global_gm_planting_2009 .html.

4. C. James, "Global Status of Commercialized Biotech/GM Crops: 2010," International Service for the Acquisition of Agri-Biotech Applications (ISAAA) Briefs no. 42, 2010.

5. S. O. Duke, "Why Have No New Herbicide Modes of Action Appeared in Recent Years?" *Pest Management Science* 68 (2012): 505–12.

6. J. K. Norsworthy et al., "Reducing the Risks of Herbicide Resistance: Best Management Practices and Recommendations," *Weed Science* 60 (2012): 31–62.

7. Ibid.

8. The United States Department of Agriculture gave approval to Dow Agro-Sciences to market 2,4-D–resistant corn and soy in January 2014.

9. B. E. Tabashnik et al., "Insect Resistance to Bt Crops: Evidence Versus Theory," *Nature Biotechnology* 26 (2008): 199–202.

10. C. M. Benbrook, "Impacts of Genetically Engineered Crops on Pesticide Use in the U.S.—The First Sixteen Years," *Environmental Sciences Europe* 24 (2012), www.enveurope.com/content/24/1/24.

11. S. Wang, D. R. Just and P. Pinstrup-Andersen, "Bt-Cotton and Secondary Pests," *International Journal of Biotechnology* 10 (2008): 113–21; Z. J. Wang et al., "Bt Cotton in China: Are Secondary

Insect Infestations Offsetting the Benefits in Farmer Fields?" *Agricultural Sciences in China* 8 (2009): 83–90.

12. See, for example: J. R. Reichman et al., "Establishment of Transgenic Herbicide-Resistant Creeping Bentgrass (*Agrostis stolonifera* L.) in Nonagronomic Habitats," *Molecular Ecology* 15 (2006): 4243–55; A. Légère, "Risks and Consequences of Gene Flow from Herbicide-Resistant Crops: Canola (*Brassica napus* L.) as a Case Study," *Pest Management Science* 61 (2005): 292–300.

13. I. Heap, "International Survey of Herbicide Resistant Weeds," Auguest 1, 2014, www.weedscience.com/summary/home.aspx.

14. S. O. Duke and S. B. Powles, "Glyphosate: A Once-in-a-Century Herbicide," *Pest Management Science* 64 (2008): 319–25; Duke, "Why Have No New Herbicide Modes of Action Appeared in Recent Years?"

15. B. R. Lu, "Transgene Escape from GM Crops and Potential Biosafety Consequences: An Environmental Perspective," *Collection of Biosafety Reviews* 4 (2008): 66–141.

16. Ibid.

17. F. Husnik et al., "Horizontal Gene Transfer from Diverse Bacteria to an Insect Genome Enables a Tripartite Nested Mealybug Symbiosis," *Cell* 153 (2012): 1567–78.

18. G. Shadle et al., "Down-Regulation of Hydroxycinnamoyl CoA: Shikimate Hydroxycinnamoyl Transferase in Transgenic Alfalfa Affects Lignification, Development and Forage Quality," *Phytochemistry* 68 (2007): 1521–29.

19. USDA ERS.

20. R. Painter, *Insect Resistance in Crop Plants* (Lawrence, KS: University of Kansas Press, 1951).

21. A. Wada-Katsumata, J. Silverman, and C. Schal, "Changes in Taste Neurons Support the Emergence of an Adaptive Behavior in Cockroaches," *Science* 340 (2013): 972–75.

22. Glyphosate-resistant Palmer amaranth (pigweed) was reported in cotton fields in Georgia in 2005 and is now widespread in the southeastern United States in cotton, soybean, and corn fields. Amaranth is a tall weed that cross-pollinates and is capable of producing tens of thousands of seeds per plant. The gene for resistance moves from one plant to the next via wind-blown pollen. In these areas, resis-

tant Palmer amaranth has essentially eliminated glyphosate from the ranks of useful herbicides, and some populations of Palmer amaranth are now resistant to at least three herbicide MOA.

Chapter 11

1.   In ecology, *irruptions* are sudden, rapid increases in population size.
2.   See again the reference to Albert Einstein at the beginning of chapter 7.

Chapter 12

1.   J. K. Norsworthy et al., "Reducing the Risks of Herbicide Resistance: Best Management Practices and Recommendations," *Weed Science* 60 (2012): 31–62.
2.   S. O. Duke and S. B. Powles, "Glyphosate: A Once-in-a-Century Herbicide," *Pest Management Science* 64 (2008): 319–25.
3.   Even Rachel Carson (*Silent Spring*, 1964) suggested the general rule: "'Spray as little as you possibly can' rather than 'Spray to the limit of your capacity.'"
4.   See: Phillips McDougall (company website), www.phillips mcdougall.com/ for estimates of time to market and costs of development; see also: S. O. Duke, "Why Have No New Herbicide Modes of Action Appeared in Recent Years?" *Pest Management Science* 68 (2012): 505–12.
5.   Insecticide Resistance Action Committee International MoA Working Group, IRAC website, 2012, www.irac-online.org/.
6.   Weed Science Society of America, "Summary of Herbicide Mechanism of Action," 2011, http://wssa.net/wp-content/uploads/WSSA -Mechanism-of-Action.pdf. WSSA and HRAC lists of MOA are equivalent.
7.   G. B. Frisvold, T. M. Hurley, and P. D. Mitchell, "Adoption of Best Management Practices to Control Weed Resistance by Corn, Cotton, and Soybean Growers," *Agbioforum* 12 (2009): 370–81.
8.   See, for example: S. Ossowski et al., "The Rate and Molecular Spectrum of Spontaneous Mutations in *Arabidopsis thaliana*," *Science* 327 (2010): 92–94. This paper estimates that mutations are inherited at the rate of two per individual for this species.

9. See, for example: University of California Agriculture and Natural Resources Statewide Integrated Pest Management Program, "UC IPM Pest Management Guidelines: Cotton," Publication 3444, 2013, www.ipm.ucdavis.edu/PMG/selectnewpest.cotton. html: "Using selective insecticides and miticides to kill the target pest without killing natural enemies helps maximize as well as integrate chemical and biological controls."

10. See: Phillips McDougall company website (www.phillips mcdougall.com/) for further information regarding the challenges and costs of bringing new agricultural chemical products to market.

Chapter 13

1. D. Q. Innis, *Intercropping and the Scientific Basis of Traditional Agriculture* (London, UK: Intermediate Technology Publications, 1997).

2. L. S. Cordell, "Late Anasazi Farming and Hunting Strategies: One Example of a Problem in Congruence," *American Antiquity* 42 (1977): 449–61.

3. W. G. Kirk, "Swine Production in the Southeast," *Journal of Animal Science* 1936b (1936): 103–6; L. L. Stephan, "Peanut Production in Southeastern United States," *Economic Geography* 21 (1945): 183–91.

4. R. Gregory and H. Guttmann, "The Ricefield Catch and Rural Food Security," pp. 1–14 in *Rural Aquaculture*, ed. P. Edwards, D. C. Little, and H. Demaine (Wallingford, UK: CABI Publishing, 1982).

5. K. Ruddle and G. Zhong, *Integrated Agriculture-Aquaculture in South China: The Dike-Pond System of the Zhujiang Delta* (Cambridge, UK: Cambridge University Press, 1988).

6. S. H. Sharaf El Din, "Effect of the Aswan High Dam on the Nile Flood and on the Estuarine and Coastal Circulation Pattern along the Mediterranean Egyptian Coast," *Limnology and Oceanography* 22 (1977): 194–207; S. Postel, *Pillar of Sand: Can the Irrigation Miracle Last?* (New York: W. W. Norton, 1999).

7. Ibid.

8. N. Myers, "Environment and Security," *Foreign Policy* 74 (1989): 23–41.

9.   M. Fukuoka, *The One-Straw Revolution* (Emmaus, PA: Rodale Press, 1978).

10.   P. Opfer and D. McGrath, "Cabbage Aphid and Green Peach Aphid," 2014 http://horticulture.oregonstate.edu/content/cabbage-aphid-green-peach-aphid; University of Florida, Department of Entomology and Nematology, "Featured Creatures: Green Peach Aphid," http://entnemdept.ufl.edu/creatures/veg/aphid/green_peach_aphid.htm#host (updated October 2005).

11.   It is worth noting that integrated pest management (IPM) programs recognize the utility of predator species and recommend their protection, but often without recommendations for managing their habitat requirements. For example, the University of California IPM recommendations for aphids in pear orchards (updated July 2013) acknowledges a range of aphid predators but suggests the reduction of aphid refuges rather than the development of predator refuges. See: www.ipm.ucdavis.edu/PMG/PESTNOTES/pn7404.html.

12.   There is strong evidence for this concept. See, for example: W. E. Snyder et al., "Predator Biodiversity Strengthens Herbivore Suppression," *Ecology Letters* 9 (2006): 789–96.

Chapter 14

1.   R. H. Painter, *Insect Resistance in Crop Plants* (Lawrence, KS: University Press of Kansas, 1951).

2.   Ibid., viii.

3.   I. Heap, "International Survey of Herbicide Resistant Weeds," August 1, 2014, www.weedscience.org/summary/Country.aspx.

4.   S. B. Powles and J. A. M. Holtum, "Herbicide-Resistant Weeds in Australia," *Proceedings of the 9th Australian Weeds Conference*, Adelaide, South Australia, August 6–10, 1990, 185–93.

5.   E. Stokstad, "The War Against Weeds Down Under," *Science* 341 (2013): 734–36.

6.   It is important to recognize that efforts in one industry can be compromised by efforts in another. In Australia, ryegrass seed production is an important component of the sheep and cattle industry, and active research and development programs continue to produce better and stronger ryegrass strains for animal forage. That industry battles growing disease and insect herbivory, and is now incorporat-

ing more biotechnology in the production of ryegrass. Of course, those efforts may contribute to the ongoing problems experienced by industries, such as the wheat industry, that view ryegrass as a major problematic weed.

7. L. White Jr., "The Historical Roots of Our Ecological Crisis," *Science* 155 (1967): 1203–7.

8. W. Jackson, *New Roots for Agriculture* (Lincoln, NE: University of Nebraska Press, 1980). Several subsequent books and collections of essays by Jackson address the multitude of topics contained in this book.

9. J. A. Dewar, "Perennial Polyculture Farming: Seeds of Another Agricultural Revolution?" publication of the Rand Corporation, Pardee Center, 2007, www.rand.org/content/dam/rand/pubs/.../2007/RAND_OP179.pdf.

10. Similar production has also been demonstrated at the Rodale Institute and the Woody Perennial Polyculture Research Site at the University of Illinois.

11. Ray's "Soil Health Page": (http://vimeo.com/channels/raythesoilguy), his "Soil Stories" (www.nrcs.usda.gov/wps/portal/nrcs/detail/national/technical/?cid = STELPRDB1043490), and his "Dirt Diaries" (http://dirtdiaries.com/author/buz/) are examples of how these soil scientists are attempting to make informative and accessible videos about the importance of soil and the maintenance of its health and structure.

12. University of California Agriculture and Natural Resources Statewide Integrated Pest Management Program, "What Is Integrated Pest Management (IPM)?" UC IPM Online, www.ipm.ucdavis.edu/GENERAL/whatisipm.html (revised July 2014).

13. University of California Agriculture and Natural Resources, "UC IPM Pest Management Guidelines: Cotton," Publication 3444 (revised May 2013).

14. Ibid., 13.

15. Ibid.

Chapter 15

1. K. Popper, *Conjectures and Refutations: The Growth of Scientific Knowledge* (London: Routledge, 1963).

2.	T. Kuhn, *The Structure of Scientific Revolutions* (Chicago: University of Chicago Press, 1962).

3.	This difficulty is related to the language we use. If we believe we can "eradicate pests" and "control nature" and that we can find technological "solutions" to natural "problems," we are caught in a trap created by our language. The cognitive confusion is perhaps best described by the concept of *linguistic relativity*, which holds that the language a culture uses will determine how that culture perceives and interprets the world. In our case, we continue to pursue solutions that are at odds with the rules of biological reality. Until a new set of rules is adopted, the appropriate solutions will continue to elude us.

INDEX

Page numbers followed by "b" indicate information in boxes.